IMMUNOPATHOLOGY, DIAGNOSIS, AND TREATMENT OF HPV-INDUCED MALIGNANCIES

Developments in Microbiology

IMMUNOPATHOLOGY, DIAGNOSIS, AND TREATMENT OF HPV-INDUCED MALIGNANCIES

Edited by

PRASHANT KHARE

Department of Microbiology, All India Institute of Medical Sciences Bhopal, Bhopal, Madhya Pradesh, India

ASHISH JAIN

Department of Microbiology, Smt. Chandibai Himathmal Mansukhani College (affiliated with the University of Mumbai), Ulhasnagar, Maharashtra, India

ELSEVIER

ACADEMIC PRESS

An imprint of Elsevier

Academic Press is an imprint of Elsevier
125 London Wall, London EC2Y 5AS, United Kingdom
525 B Street, Suite 1650, San Diego, CA 92101, United States
50 Hampshire Street, 5th Floor, Cambridge, MA 02139, United States
The Boulevard, Langford Lane, Kidlington, Oxford OX5 1GB, United Kingdom

ISBN 978-0-323-91797-1

For information on all Academic Press publications
visit our website at https://www.elsevier.com/books-and-journals

Publisher: Stacy Masucci
Acquisitions Editor: Linda Versteeg-Buschman
Editorial Project Manager: Tim Eslava
Production Project Manager: Niranjan Bhaskaran
Cover Designer: Greg Harris

Typeset by STRAIVE, India

Contents

Contributors *ix*
Preface *xi*

1. An overview of HPV: Causes, symptoms, and clinical manifestations **1**

Anuj Tripathi and Utkarsha Sahu

1. Introduction 1
2. Structure and biology of HPV 3
3. Causes of HPV infection 4
4. HPV symptoms 5
5. HPV diagnosis 5
6. Clinical manifestation of HPV infections 7
7. Prevention of HPV infection 10
8. Wart treatment 11
9. HPV vaccines 11
10. Conclusion 12

References 13

2. Pathogenesis and immune response against HPV infection **21**

Utkarsha Sahu and Prashant Khare

1. Introduction 21
2. Pathogenesis and immune response associated with HPV infection 22
3. Evasion of host immune response 30
4. Conclusion 32

References 34

3. Malignancies associated with HPV **43**

Prashant Kumar Tiwari, Divya Kushwaha, and Ambuj Kumar Kushwaha

1. Introduction 43
2. Malignancies associated with HPV 46
3. Treatment of HPV-associated malignancies 55
4. Conclusion 56

Acknowledgment 56

References 57

4. Immunological aspects of HPV/HIV co-infection and risk of gynecological diseases **65**

Amrita Srivastava, Priyanka Namdev, and Rajaneesh Anupam

1. Introduction 65
2. Immunological aspects of HPV infection 68
3. The impact of immunological aspects of HPV on HIV infection 70
4. HPV-HIV co-infections and gynecological disorders 72
5. Conclusion 75

References 76

5. Cervicovaginal microbiota and HPV-induced cervical cancer **81**

Ashish Jain, Swadesh Kishore Shrivastava, and Linju Joy

1. Introduction 81
2. Human papillomavirus (HPV) 82
3. Cervicovaginal microbiota (CVM) 83
4. CVM, sexually transmitted infections (STIs), and CC 86
5. Restoration of CVM to prevent CC 90
6. Conclusion 90

References 91

6. Current diagnostic tools for HPV **99**

Aneesh Sethu Madhavan, Deepti Joshi, Tanya Sharma, Shashwati Nema, and Debasis Biswas

1. Introduction 99
2. Diagnostic techniques for HPV infection 100
3. Screening strategies 112
4. Conclusion 115

References 115

7. Animal models for the discovery of therapeutic and prophylactic measure against HPV-induced malignancies **119**

Shobha Upreti and Mukesh Samant

1. Introduction 119
2. Animal models for therapeutic and prophylactic measures against HPV 121
3. Conclusion 125

References 126

8. Potential drug targets against HPV and repurposing of existing drugs **129**

Utkarsha Sahu, Prashant Khare, and Ashvini Yadav

1. Introduction 129
2. Potential drug targets against HPV 130
3. Repurposing of existing drugs against HPV 137
4. Conclusion 141

References 143

9. Recent advances and strategies in vaccine development against HPV **149**

Anoop Kumar and Praveen Rai

1. Introduction 149
2. Requirement and type of HPV vaccine 151
3. Candidate for the HPV vaccine development 152
4. Licensed vaccines against HPV 153
5. Development of the future HPV vaccine 153
6. The future of HPV vaccination 160
7. Conclusion 160

References 160

10. Available immunotherapies and future opportunities to prevent HPV-associated cancers **165**

Rohit Kumar Tiwari, Afza Ahmad, Irfan Ahmad Ansari, and Preeti Bajpai

1. Introduction 165
2. Present therapeutic options for HPV-associated cancers 168
3. Immunotherapies for HPV-associated cancers 169
4. CC 170
5. HNSCC 181
6. Oral cancer 190
7. Conclusion 192

References 192

Index *205*

Contributors

Afza Ahmad
Department of Biosciences, Integral University, Lucknow, Uttar Pradesh, India

Irfan Ahmad Ansari
Department of Biosciences, Integral University, Lucknow, Uttar Pradesh, India

Rajaneesh Anupam
Department of Biotechnology, Dr. Harisingh Gour Vishwavidyalaya (A Central University), Sagar, Madhya Pradesh, India

Preeti Bajpai
Department of Zoology, Mahatma Gandhi Central University, Motihari, Bihar, India

Debasis Biswas
Department of Microbiology, All India Institute of Medical Sciences Bhopal, Bhopal, Madhya Pradesh, India

Ashish Jain
Smt. Chandibai Himathmal Mansukhani College, Ulhasnagar-3, Affiliated to University of Mumbai (Maharashtra), Mumbai, Maharashtra, India

Deepti Joshi
Department of Pathology & Laboratory Medicine, All India Institute of Medical Sciences Bhopal, Bhopal, Madhya Pradesh, India

Linju Joy
Smt. Chandibai Himathmal Mansukhani College, Ulhasnagar-3, Affiliated to University of Mumbai (Maharashtra), Mumbai, Maharashtra, India

Prashant Khare
Department of Microbiology, All India Institute of Medical Sciences, Bhopal, Madhya Pradesh, India

Anoop Kumar
National Institute of Biologicals (NIB), Ministry of Health & Family Welfare, Government of India, Noida, Uttar Pradesh, India

Ambuj Kumar Kushwaha
School of Biotechnology, Institute of Science, Banaras Hindu University, Varanasi, UP, India

Divya Kushwaha
Department of Chemistry, MMV, Banaras Hindu University, Varanasi, UP, India

Aneesh Sethu Madhavan
Department of Microbiology, All India Institute of Medical Sciences Bhopal, Bhopal, Madhya Pradesh, India

Priyanka Namdev
Department of Biotechnology, Dr. Harisingh Gour Vishwavidyalaya (A Central University), Sagar, Madhya Pradesh, India; Centers for Disease Control and Prevention, Atlanta, GA, United States

Shashwati Nema
Department of Microbiology, All India Institute of Medical Sciences Bhopal, Bhopal, Madhya Pradesh, India

Praveen Rai
Division of Infectious Diseases, Nitte (Deemed to be University), Nitte University Centre for Science Education & Research (NUCSER), Deralakatte, Karnataka, India

Utkarsha Sahu
Department of Microbiology, All India Institute of Medical Sciences, Bhopal, Madhya Pradesh, India

Mukesh Samant
Cell and Molecular Biology Laboratory, Department of Zoology, Kumaun University, Almora, Uttarakhand, India

Tanya Sharma
Department of Pathology & Laboratory Medicine, All India Institute of Medical Sciences Bhopal, Bhopal, Madhya Pradesh, India

Swadesh Kishore Shrivastava
Omisure Technologies, Bhopal, Madhya Pradesh, India

Amrita Srivastava
Department of Biotechnology, Dr. Harisingh Gour Vishwavidyalaya (A Central University), Sagar, Madhya Pradesh, India

Prashant Kumar Tiwari
School of Biotechnology, Institute of Science, Banaras Hindu University, Varanasi, UP, India

Rohit Kumar Tiwari
Department of Biosciences, Integral University, Lucknow, Uttar Pradesh, India

Anuj Tripathi
Department of Microbiology, Immunology, and Physiology, Meharry Medical College, Nashville, TN, United States

Shobha Upreti
Cell and Molecular Biology Laboratory, Department of Zoology, Kumaun University, Almora, Uttarakhand, India

Ashvini Yadav
Department of Microbiology, All India Institute of Medical Sciences, Bhopal, Madhya Pradesh, India

Preface

This book is meant to offer a complete overview of human papillomavirus (HPV) from discovery to the present day and its association with cervical malignancies. HPV is a group of more than 200 types of viruses and the most common viral infection of the reproductive tract. HPV is associated with cancers of the cervix, vulva, vagina, anus, penis, and oral cavity. Persistent infection of high-risk HPV (HR-HPV) can cause cervical cancer, the fourth most common cancer among women, with annually 570,000 new cases and 311,000 deaths reported globally predominantly in low- and middle-income nations. The book starts with an overview of HPV and provides a comprehensive update covering its pathogenesis, clinical manifestation, and current diagnostic tools. Further, it focuses on HPV-associated malignancies, host immune response against infection, animal models used in HPV research, and role of cervicovaginal microbiota and other sexually transmitted infections in disease progression and prevention. The book also elaborates available immunotherapies, novel targets on virus for drug development, different strategies for vaccine development, and future direction in HPV-induced cancer research. It is a useful reference for researchers, scientists, oncologists, academicians, undergraduate and postgraduate students and for anyone working in the HPV field to gain an understanding of different aspects of HPV-induced malignancies.

CHAPTER 1

An overview of HPV: Causes, symptoms, and clinical manifestations

Anuj Tripathi[a] and Utkarsha Sahu[b]
[a]Department of Microbiology, Immunology, and Physiology, Meharry Medical College, Nashville, TN, United States
[b]Department of Microbiology, All India Institute of Medical Sciences, Bhopal, Madhya Pradesh, India

1. Introduction

Papillomaviruses are a common group of viruses, including small, nonenveloped viruses that infect and induce cellular proliferation in cutaneous and mucosal epithelia of higher vertebrates [1,2]. Globally, about 570,000 cases per year in women and 60,000 cases in men are attributed to the human papillomavirus (HPV). Worldwide, 40% of all types of cancers, including approximately 2% from developed countries and 8% from developing countries, are caused by HPV infection. World Health Organization (WHO) report indicated that HPV causes 610,000 cancer cases and 250,000 deaths every year [3–6]. Papillomaviruses are highly species-specific and infect many vertebrates. All the papillomaviruses are approximately similar in genome organization and physical structure. The virus genome is composed of approximately 8 kb, double-stranded, circular DNA genome enclosed within a 55-nm-diameter capsid [1,7]. Based on their association with infection, cancer, and precursor lesions, HPVs can also be classified into two groups: high-risk (HR-HPV) and low-risk (LR-HPV) oncogenic types. In LR-HPV types, such as HPV6 and HPV11, infection can develop common genital warts or benign lesions, while in HR-HPV types, like HPV16 and HPV18, infection is associated with cervical cancer (CC) [8–11]. HR-HPV types are the causes of 40% of oral cancers and can induce anal, penile, vulvar, mouth, and throat carcinomas [12].

HPVs include over 100 different genotypes that infect basal epithelial cells of the skin or inner lining tissues. Some of the HPVs are responsible for cutaneous warts, while one subset of HPVs is found in the skin lesions

Immunopathology, Diagnosis, and Treatment of HPV-induced Malignancies
https://doi.org/10.1016/B978-0-323-91797-1.00003-X

associated with cutaneous squamous cell carcinomas [13,14]. Some HPVs are associated with infections that can cause benign, premalignant, and malignant lesions not only of the genital tract and anus but also of the oropharynx [14,15]. Out of 100 types of HPV, approximately 14 are cancer-causing in humans [8,16,17]. HPV infection is usually a transient dermatologic infection. Genital HPV infection is transmitted through sexual contact very shortly and is one of the most common sexually transmitted infections (STIs) in both men and women [18,19]. The HPV structural and nonstructural proteins play a crucial role in the attachment, entry, and replication of HPV. In addition, these proteins also help to avoid recognition by the host immune system [20]. During the initial phase of HPV infection, the virus maintains a low copy number to avoid immune recognition. Later, the rate of replication increases as the viral proteins hijack the cell cycle regulatory proteins and downregulated host immune factors [21]. The viral infection triggers a dysregulated inflammatory response that leads to viral persistence and tumor progression [22,23].

Once the HPV infection is established, it leads to different clinical manifestations such as warts, cervical dysplasia, or cervical intraepithelial neoplasia grade 1 (CIN 1). Generally, these lesions are self-limited by the host immune system within a year [24,25]. The precise immune mechanism involved in the clearance of HPV infection is still not fully explored. However, the major immune factors activated during HPV infection include tumor necrosis factor (TNF), interferon (IFN), transforming growth factor-β (TGF-β), Langerhans cells, and dendritic cells [25]. The role of TGF-β and TNF is not clear. They are reported to play both pro- and antiviral roles depending upon the experimental conditions [26–31]. Some of the HPV-induced clinical manifestations are not cleared by the host immune response, and their persistence can lead to malignant progressions such as squamous cell carcinoma and cervical adenocarcinoma [32]. Although it was initially reported that HPV primarily targets the women's genital tract, it has also been associated with malignancies such as penile cancer, mouth cancer, and throat cancer [33–39].

Even after the past 50 years of HPV research, we still lack a complete understanding of HPVs and associated pathogenesis patterns. Also, most of the work on HPV biology is associated with HR-HPV strains. Our understanding of LR-HPV strains remains elusive. The early diagnosis of HPV can help in limiting the infection and prevent the associated clinical manifestations. Different diagnostic techniques used against HPV include Pap smear, nucleic acid detection, and colposcopy. Once the HPV is detected, the general treatment against the associated warts or lesions is administered. This includes different physical and chemical treatments [40–44]. The vaccines against HPV are mostly prophylactic and lack therapeutic applications. The development of therapeutic vaccines is a need in the present scenario. In this chapter, we have discussed an overview of the causes, symptoms, and clinical manifestations of HPV infection, and a better knowledge of different factors associated with HPV infection would pave the way for the development of effective prophylactic and therapeutic measures against HPVs.

2. Structure and biology of HPV

HPV infects cutaneous and mucosal epithelial cells in humans. The 8 kb viral genome encodes eight open reading frames (ORFs) [45]. The structure and function of the HPV genome are conserved and have three major regions:

- **(i)** Long control upstream regulatory region (URR), located between the L1 and E6 open reading frames (ORFs). These regions have a promoter and regulatory elements that are involved in viral DNA replication and transcription.
- **(ii)** Early regions encode the E1, E2, E4, E5, E6, and E7 proteins. These proteins are involved in viral gene replication, expression, and survival.
- **(iii)** The late region encodes the structural proteins (L1 and L2).

This organization is shared among all α-HPVs. Certain HPV subtypes may lack some ORF because only four ORFs (those of E1, E2, L1, and L2) are necessary to ensure viral replication and infection [46]. All papillomaviruses possess a nonenveloped icosahedral capsid. The virus contains 360 molecules of L1 protein that forms 72 capsomeres [47–50]. HPV infection occurs at the basal layer of stratified squamous epithelial cells. Infection stimulates cell proliferation in the epithelium, and the cells undergo different changes from benign hyperplasia to dysplasia to invasive and later to carcinoma. In general cases, HPV infection is cleared by the host immune response, but some strains of HPV may develop carcinogenic warts and lesions [20] (Fig. 1).

The replication cycle of HPV can be divided into three different stages, namely establishment, maintenance, and amplification [51]. The viruses enter the host epithelium through microlesions and infect the basal epithelial cells [52]. The virus interacts with the host extracellular matrix via the capsid proteins [53,54]. This interaction results in the internalization of the virus particle inside the cell. Inside the target cell, the viral DNA is uncoated and

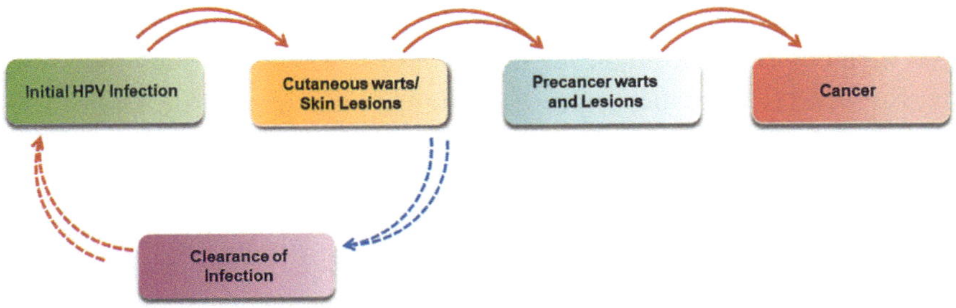

Fig. 1 Schematic representation of HPV pathogenesis.

transported to the nucleus, where it creates multiple copies [52]. E1 and E2 proteins are expressed in the HPV-infected basal layer of epithelium and regulate early viral DNA transcription. Approximately 50–100 HPV DNA copies per cell are stably maintained by E1 and E2 viral proteins in the undifferentiated basal cell [20,51]. In later stages, the E6 and E7 proteins are expressed and nullify the cell cycle regulation leading to continuous cell proliferation. HR E6 binds to the p53 tumor suppressor protein and leads to the rapid turnover of p53 [55,56]. E7 binds to the tumor suppressor retinoblastoma (Rb) family proteins and other proteins involved in cell cycle regulation [57,58]. Altogether, these events result in failure of cell cycle regulation and malignant transformation of infected host epithelial cells. Viral genomes are divided into HPV-infected basal cells. Infected basal cell leaves the basal layer and undergoes differentiation, due to action of the E7 protein [59]. Infected cells reenter in the S phase and convert into highly differentiated cells that have activated cellular replication factors required for viral replication. The E4 and E5 proteins are involved in the regulation of late viral functions [60]. The L1 and L2 proteins are assembled and form icosahedral capsids in later stages. L1 protein is the major capsid protein and is highly immunogenic, which is expressed in the last to avoid immune surveillance [61]. The mature viruses are assembled and released from the uppermost layers of the epithelium [61].

3. Causes of HPV infection

HPV infection is contagious; i.e., they are transmitted from one infected person to another via skin-to-skin contact [62]. The virus usually enters the human body, through a small cut, tear, or abrasion in the skin [15,19,63]. People get genital HPV infections through direct sexual intercourse, including vaginal, anal, and oral sex, and other skin-to-skin contacts in the genital region. Baby can get HPV infection through genital warts from HPV-infected mother [64].

4. HPV symptoms

HPV symptoms may be seen years after the initial infection. Early HPV infection does not cause any noticeable symptoms or health problems. The symptomatic HPV infections are marked by the presence of warts. Warts are very contagious and can spread through direct contact with a wart or indirectly when someone touches things that already have contact with a wart [63]. Some types of viruses cause warts to form, and in 90% of cases, they go away within two years. However, prolonged infection with HPV in the skin can cause serious health problems, including genital warts and throat warts. These types of warts can increase the risk of cancer in the body at later stages [16,65]. HPV warts can be categorized into two types, namely common warts and genital warts.

4.1 Common warts

HPV infection generally causes common warts, also known as flat warts or plantar warts. Common warts are bumpy, rough, sometimes hard, flat-topped, and slightly raised that tend to be formed on the hands, feet, fingers, and face or neck [14,18,66].

4.2 Genital warts

HPV infection with genital organs causes genital warts. They can be one small skin bump or cluster of bumps. These warts can cause burning, itching, and other symptoms. Genital warts can range in size and appearance, and they may be large or small; flat or cauliflower-shaped; white, pinkish-red, or purplish-brown colored. They can be formed on the vulva, cervix, penis, and anus [2,67]. The asymptomatic patients are silent carriers of infection. Asymptomatic shedding of HPV/DNA from the peri-anal region is very common. Especially, the transmission rate in patients with acute immunodeficiency syndrome (AIDS) can be up to 78.9% [62].

5. HPV diagnosis

Accurate diagnostics of infection at the right time is a golden chance to cure the disease and save the life. In recent eras, advanced diagnostic techniques have been available to identify and cure many infectious diseases [68].

Regular screenings with a physical examination of any warts or lesions can help diagnose HPV-related health problems earlier. Most warts are benign, but any of them can be converted into malignancies. Warts can occur on the skin, vagina, cervix, anus, and oropharyngeal areas [16,69]. The conventional methods for HPV diagnosis were less efficient and, thus, not suitable for proper HPV detection. These methods included cell culture electron microscopy and immunological methods [62]. Currently, more effective diagnostic techniques such as colposcopy and acetic acid test, biopsy, nucleic acid detection using PCR, in situ hybridization, Southern blot hybridization, and pap smear [62] are employed for HPV detection. Serological tests are not very promising for HPV infection due to weak host serological response. US preventive services task force (USPSTF) suggested that women aged between 21 and 29 should have a Pap test every three years [70]. Women aged between 30 and 65 should have a Pap test every three years with HPV test every five years. Abnormal Pap test in less than 30 years of age also should be confirmed through HPV test with the recommendation of doctor or gynecologist [65,70].

5.1 HPV tests in women

Regular Pap tests help to identify abnormal cells in women. According to guidelines of United States, at least 14 strains of HPVs are directly related to cancer, and confirmation of these strains should monitor cervical changes in patients [70–72]. Another technique includes the colposcopy of patients. Colposcopy allows biopsy targeted to the atypical areas. The biopsy of atypical areas is a crucial part of colposcopy since the treatment depends on the severity displayed by the biopsy sample. If the results suggest precancer (dysplasia) or malignancy, then treatment is a must. The dysplasia might be minor, temperate, or severe [62]. In addition, nucleic acid tests are also performed to diagnose the presence of HPV. These include polymerase gene reaction (PCR) and in situ hybridization [62].

5.2 HPV tests in men

Currently, there is no approved test for HPV diagnosis in men. The HPV nucleic acid test is only available for diagnosing HPV in women. Routine screening to check for HPV-related disease in men is not currently recommended by CDC [73,74]. Few doctors might perform an anal Pap test for men with an augmented risk of developing anal cancer. This includes men with HIV and men involved in anal sex with men [73,75].

6. Clinical manifestation of HPV infections

Globally, 20%–25% of all cancer cases are related to any infectious agents, and the virus alone is responsible for 12% of all human cancers [76,77]. HPV has a direct relation with approximately 30% of all infectious agent-related cancers [21]. In these cases, most deaths came from low- and middle-income countries. Now, HPV infection is gripping developed countries, and about 6.2 million people in the United States are annually infected with HPV [3,78]. The clinical manifestations of HPV during the initial phase are warts, CIN, lesions, etc. However, if left untreated, they may lead to CC, vaginal cancer, vulva cancer, anal cancer, penile cancer, and oropharyngeal cancer. Every year, HPV infection has arrested large number of populations in India [33]. The HPV-related cancer distribution in the Indian population is shown in Fig. 2.

Warts and lesions are generally cleared with effective treatment. However, their malignant progression may lead to different types of malignancies such as CC, vaginal cancer, penile cancer, vulvar cancer, mouth cancer, and throat cancer (Fig. 3).

6.1 CC

WHO estimated 27 million new cases of CC and 17 million deaths by the disease in 2030 [6]. Women of sub-Saharan Africa, Eastern Europe, and Latin America have a high number of infections and increased cancer

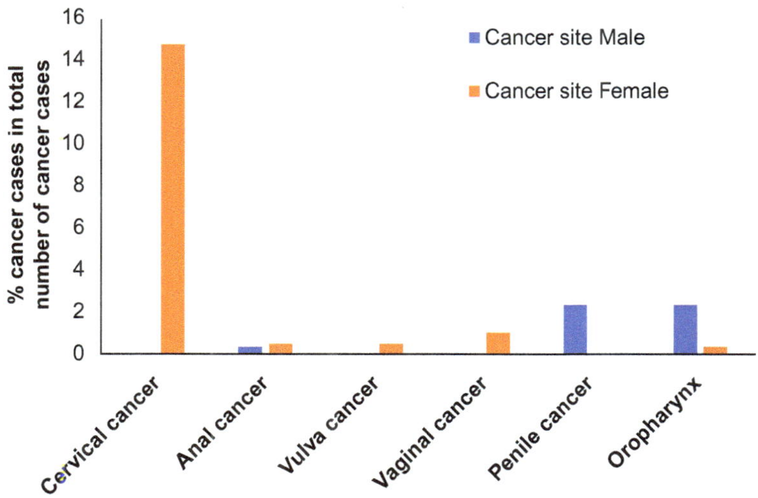

Fig. 2 HPV-related cancer distribution in the Indian population. [Different types of HPV cancer distribution in Indian male and female population]. *(Data from HPV Information Centre reports, HPV CC in female prominent in Indian population).*

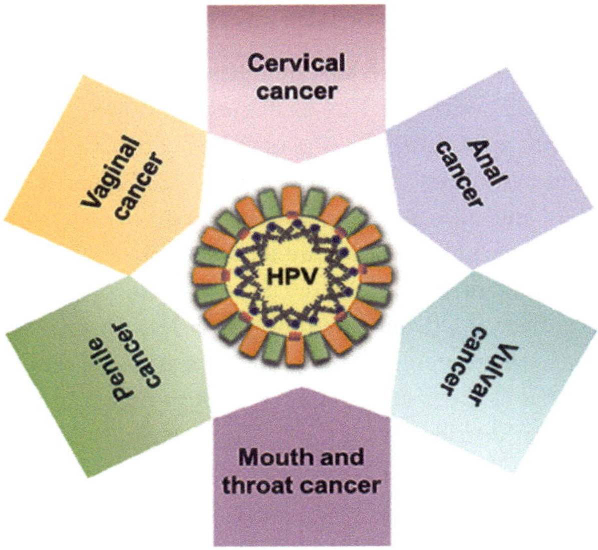

Fig. 3 Different malignancies caused by HPV in humans.

incidence in developing countries [79]. Only in India, every year, 96,922 women are diagnosed with CC, and 60,078 die from the disease. Currently, CC is the second most frequent cancer among women in India [33].

Warts in the cervix start appearing 2–3 months after infection. The CC from HPV infection has taken a long time, with a premalignant lesion development (with 3–5 years), to cancer development (10–30 years) [80]. HPV is sexually transmitted in women during intercourse [79] and directly infects the squamocolumnar junction cell [81,82]. Sexually active ~80% of women get HPV infection during their lifetime [81]. Due to the lack of an effective immune response, about 10% of infected women have a risk of developing cancer. Early detection and treatment of abnormal cervical tissues can prevent CC. Cancer is always considered a multifactorial disease, so other environmental factors like smoking, alcohol consumption, long-term contraceptive hormone, and coinfection of HIV increase the risk of CC [70,79,83].

6.2 Vaginal cancer

Vaginal cancer is rare and is more common in postmenopausal women, but the number of cases rises in young women with a high prevalence of HPV infection [84]. Current data suggest that HPV infection may be correlated with approximately 78% of vaginal cancers worldwide [34]. In general, vaginal warts contain HPV in the precancer stage for many years before converting into the cancer stage. In very few cases, Pap tests recognized vaginal cancer in precancer stages [85]. Since vaginal carcinoma is a rare entity, cure and treatment depend on the stage of disease and the time of diagnosis [86]. Available treatment of vaginal cancers involves surgical vaginectomy with chemotherapy and radiotherapy [87].

6.3 Vulvar cancer

The average age of women diagnosed with vulvar cancer is 65–70 years and shares 5% of all gynecologic cancers, but the frequency of vulvar cancer in young women has increased significantly with HPV infection [35–37]. The vulva is the external part of the female genitals, including the opening of the vagina, the labia majora, the labia minora, and the clitoris. Studies show that vulvar cancer is associated with HPV virus subtypes 16 and 18 [88,89]. In the early stage, vulvar cancer patients should undergo a surgical excision including adjuvant chemotherapy treatment, but advanced-stage disease has no standard recommended treatment strategy [37,90]. Recent data show some hope with neoadjuvant chemotherapy in the case of advanced vulvar cancer patients [91].

6.4 Penile cancer

In men, HPV is a big factor for penis cancer. Penile cancer is a rare disease, which can cause significant health issues in the initial and advanced stages. HPV infection has been directly linked to penile cancer, although exact signaling pathways and targets have not been fully elucidated [38]. Penile cancer with HPV infection occurrence is low currently in the United States but frequent in the developing countries of Latin America, Africa, and Asia [92,93]. Data show that in HIV-positive and homosexual men, HPV-related cancer of the penis is more common. Routine physical examinations work well in the early detection of penile cancer without specific tests [94,95].

6.5 Anal cancer

Anal cancer is rare, but cases have been increasing over the past some decades in both women and men. Most of the anal cancers are squamous cell cancers, and 91% of anal cancers are linked with HPV infection [39]. It is more common in the population with unsafe sex and is infected with HIV. Specific test, as well as routine screening, is not recommended in the case of anal cancer, but the expert agrees to do anal Pap cytology test for persons with a high risk of anal cancer [96,97]. Anal cancer in homosexual men is common with HPV6, HPV11, HPV16, or HPV18 subtype infections. Vaccination has become an important strategy for anal cancer prevention [98].

6.6 Mouth cancer and throat cancer

Cancers of the oral cavity and pharynx account for 3% of cancers diagnosed each year [35], and tobacco consumption, heavy alcohol drinking, and poor nutrition can cause mouth cancer and throat cancer [99]. Poor mouth hygiene and HPV infection increase the chance of oropharyngeal cancers (back of the throat, including the base of the tongue and tonsils) [100–104]. Oral sex is the driving factor behind the oropharyngeal HPV infection and transmission [105,106]. Currently, there is no standard early-stage screening test, so physical examination remains critical for determining the clinical stage of disease and planning, curing, and treating oral cavity carcinoma. Surgical removal of the primary tumor with radiation and adjuvant chemotherapy is an available treatment in the case of oral cancer and neck cancer [107].

7. Prevention of HPV infection

Some strains of HPV infection can develop cancer, and many others can cause facial and genital warts. Following suggestions can lower the risk of HPV infection [37,39,67,107].

1. HPV vaccination (Gardasil 9®) is a good way to prevent HPV-induced cancers and genital warts.
2. Avoid using towels and clothes from HPV-infected people.
3. Avoid hugging or holding the hands of infected people to lower the HPV infection.
4. Limiting the use of unhygienic public toilet seats or swimming pools is also a good way to protect from HPV infection.
5. Limiting the number of sex partners at any age can limit the risk of HPV infection.
6. Using protection (like condoms) in every sexual encounter not only reduces HPV infection, but also lowers the risk of other STD diseases and HIV infections.
7. Avoiding early age sex activity reduces the chance of HPV infection exposure.
8. Adopting a scheduled and healthy lifestyle increases the strength of the immune system and protects the body from virus attacks.

8. Wart treatment

There are physical and chemical treatments available to cure warts in the skin.

8.1 Physical treatment

Cryotherapy and surgical removal of warts are the most effective methods for the cure of wart on the face, skin, or outer genital part. In cryotherapy, liquid nitrogen is used for removing the wart. The treatment is costly but quick and safe. Another method is the surgical removal of warts by doctors, which is a cheap alternative for cryotherapy [40–42].

8.2 Chemical treatment

In most cases, warts are caused by HPV infection in the skin and are very common in people with impaired immune systems. A wart may resolve spontaneously if people have an adequate immune system, but some subtypes of HPV are associated with a risk of cancers [13,14] (4, 5). There are some chemical ointments that are effective in the treatment of wart with some limitations [42–44].

(a) Imiquimod 5% cream
Imiquimod 1%–5% creams are used to clear warts with some local irritation. Imiquimod is an immunomodulatory agent that can be used in the treatment of both common and genital warts. It stimulates the interferon-alpha and other cytokine production and increases antiviral activity in the patient's skin [108].

(b) Podofilox 0.5% gel
Podofilox 0.5% gel contains the most active compound called podophyllin and is used in the treatment of external genital warts. Podofilox is not recommended for the treatment of peri-anal, rectal, urethral, or vaginal lesions [109].

(c) Polyphenon E 15% ointment
Polyphenon E (Sinecatechins) contains a potent active compound called catechin against cancerous and virally infected cells. It is extracted from green tea leaves (*Camellia sinensis*) and has a 15% green tea polyphenols in the ointment for the treatment of genital warts [110].

9. HPV vaccines

HPV vaccines deliver robust protection against HPV infection. Vaccination is a mode of prevention, so the HPV vaccine is not used to treat HPV infections that already exist in the body [111–113]. HPV vaccination is recommended at the age of 11 or 12 years, and this vaccination offers the best protection against HPV [111,112]. Vaccination for HPV should be taken at the age of 26 years if not vaccinated earlier for HPV. Only three vaccines have been licensed in the United States for the prevention of HPV infection, Gardasil® (from Merck), Gardasil 9® (from Merck), and Cervarix® (from GSK) [114,115]. Currently, Gardasil 9® has been used as an HPV vaccine in the United States; since 2016, Gardasil 9® has effective protection against nine types of HPV infections, which include HPV types 6 and 11 (cause of 90% genital warts), types 16 and 18 (high-risk HPVs, cause of 70% of CC), types 31, 33, 45, 52, and 58 (HR-HPVs, cause of 10%–20% of CC) [17,115,116].

Currently, there is no therapeutic vaccine available against HPV. This might be due to the fact that once the HPV genome is integrated into the host genome, it is difficult for antibodies to reach the nucleus and target viral genome replication [117]. Different approaches are being done to develop different chimeric vaccines that might have therapeutic applications [118–120].

10. Conclusion

Globally, HPV infection is common, and avoidance of HPV exposure is very difficult. HPV infection is usually transmitted by contact and can cause a variety of warts and diseases. [34]. HPV infection is associated with 90% of cervical and anal cancer cases, 70% of vaginal and vulvar cancer cases, 70% of oropharyngeal cancer cases, and more than 60% of penile cancer cases [34,121]. Many strains of HPV cause genital HPV infection that can develop into a variety of anogenital diseases, including warts, lesions, and cancers. The HPV infection is directly associated with both premalignant and malignant lesions in the cervix [19,122–124]. CC is the fourth most common women's cancer around the world, with estimated ~570,000 new cases and 311,000 deaths, of which 90% occurred in developing countries [73,125]. A total of 453.02 million women at the age of 15 years and older are at risk of developing CC in India alone [33].

The cervicovaginal microbiota play a crucial role in the onset of HPV infection and associated manifestations. The imbalance in the microbiota or microbial dysbiosis can aid HPV infection and persistence by causing DNA damage and promoting a pro-inflammatory response [126]. HPV infection is also common in people with existing human immunodeficiency virus (HIV) infection. This is contributed due to reduced cell-mediated immunity in HIV-infected person that facilitates the HPV infection and persistence. Women with HPV-HIV coinfection are also more prone to suffer from invasive cervical and vulvar cancer [127].

Currently, our understanding of HPV infection and associated cancer could reach 10–30 years, but finding novel therapies targets, in case of HPV infection to cancer development, is so far. Even with the recent advances in HPV pathology and development in HPV vaccines, HPV-related cancer remains growing worldwide. HPV detection assays differ in their sensitivity and type specificity, and in many cases, there is no diagnostic technique for early detection of HPV infection and its progression in cancer. The development of detection methods for different types of HPV infection is crucial for the control of a new number of cancer cases. Thus, prevention of HPV infection is better than therapies and treatments. In the current scenario, vaccination, hygiene, limiting sexual partners, using barriers (like condoms), and male circumcision are effective prevention strategies. However, there is still a lack of therapeutic vaccines against HPV that could clear the established infection. Various in silico approaches are being carried out to design different vaccine candidates with therapeutic interventions to clear the pre-established HPV infection [118–120]. The HPV vaccination will provide a great cure of HPV-associated diseases among young males and females. It will be interesting to see, in the future, how HPV vaccination development controls HPV infection and how much HPV cancer rates decrease globally with this vaccination.

Also, identification of novel drug targets and drug discovery to clear the established HPV infection is also required. However, this is a very time-taking and costly process that could be overcome by drug repurposing approaches. Finally, in this context of public health, more translational research, funding, and collaboration are needed to increase therapeutic alternatives and curative pathway options for HPV infection.

References

[1] M.E. Scheurer, G. Tortolero-Luna, K. Adler-Storthz, Human papillomavirus infection: biology, epidemiology, and prevention, Int. J. Gynecol. Cancer 15 (5) (2005) 727–746.

[2] C. Chelimo, et al., Risk factors for and prevention of human papillomaviruses (HPV), genital warts and cervical cancer, J. Inf. Secur. 66 (3) (2013) 207–217.

[3] D.R. Lowy, HPV vaccination to prevent cervical cancer and other HPV-associated disease: from basic science to effective interventions, J. Clin. Invest. 126 (1) (2016) 5–11.

[4] M. Cullen, et al., Deep sequencing of HPV16 genomes: a new high-throughput tool for exploring the carcinogenicity and natural history of HPV16 infection, Papillomavirus Res. 1 (2015) 3–11.

[5] D. Stelzle, et al., Estimates of the global burden of cervical cancer associated with HIV, Lancet Glob. Health 9 (2) (2021) e161–e169.
[6] Report-WHO; Human Papillomavirus (HPV) and Cervical Cancer. Available from: https://www.who.int/news-room/fact-sheets/detail/human-papillomavirus-(hpv)-and-cervical-cancer.
[7] K. Munger, et al., Biological activities and molecular targets of the human papillomavirus E7 oncoprotein, Oncogene 20 (54) (2001) 7888–7898.
[8] G.M. Clifford, et al., Comparison of HPV type distribution in high-grade cervical lesions and cervical cancer: a meta-analysis, Br. J. Cancer 89 (1) (2003) 101–105.
[9] F.X. Bosch, et al., The causal relation between human papillomavirus and cervical cancer, J. Clin. Pathol. 55 (4) (2002) 244–265.
[10] F.X. Bosch, et al., Epidemiology and natural history of human papillomavirus infections and type-specific implications in cervical neoplasia, Vaccine 26 (Suppl. 10) (2008) K1–16.
[11] N. Munoz, et al., Epidemiologic classification of human papillomavirus types associated with cervical cancer, N. Engl. J. Med. 348 (6) (2003) 518–527.
[12] M. Stanley, Pathology and epidemiology of HPV infection in females, Gynecol. Oncol. 117 (2 Suppl) (2010) S5–10.
[13] N. Egawa, J. Doorbar, The low-risk papillomaviruses, Virus Res. 231 (2017) 119–127.
[14] S.V. Graham, The human papillomavirus replication cycle, and its links to cancer progression: a comprehensive review, Clin. Sci. (Lond.) 131 (17) (2017) 2201–2221.
[15] J.D. Oriel, Natural history of genital warts, Br. J. Vener. Dis. 47 (1) (1971) 1–13.
[16] L. Mariani, et al., The clinical implementation of primary HPV screening, Int. J. Gynaecol. Obstet. 136 (3) (2017) 266–271.
[17] I. Manini, E. Montomoli, Epidemiology and prevention of human papillomavirus, Ann. Ig. 30 (4 Suppl. 1) (2018) 28–32.
[18] M. Giannaki, et al., Human papillomavirus (HPV) genotyping of cutaneous warts in Greek children, Pediatr. Dermatol. 30 (6) (2013) 730–735.
[19] A.B. Moscicki, et al., Updating the natural history of human papillomavirus and anogenital cancers, Vaccine 30 (Suppl. 5) (2012) F24–F33.
[20] M.S. Longworth, L.A. Laimins, Pathogenesis of human papillomaviruses in differentiating epithelia, Microbiol. Mol. Biol. Rev. 68 (2) (2004) 362–372.
[21] I.G. Bravo, S. de Sanjose, M. Gottschling, The clinical importance of understanding the evolution of papillomaviruses, Trends Microbiol. 18 (10) (2010) 432–438.
[22] U. Sahu, et al., Interleukin-17—a multifaceted cytokine in viral infections, J. Cell. Physiol. (2021) 1–2.
[23] U. Sahu, P. Khare, Role of interleukin-17 in human papillomavirus infection and associated malignancies, Microb. Pathog. (2021) 105294.
[24] R. Höpfl, et al., Spontaneous regression of CIN and delayed-type hypersensitivity to HPV-16 oncoprotein E7, Lancet 356 (9246) (2000) 1985–1986.
[25] A.B. Jenson, R.J. Kurman, W.D. Lancaster, Tissue effects of and host response to human papillomavirus infection, Dermatol. Clin. 9 (2) (1991) 203–209.
[26] L. Braun, et al., Regulation of growth and gene expression in human papillomavirus-transformed keratinocytes by transforming growth factor-beta: implications for the control of papillomavirus infection, Mol. Carcinog. 6 (2) (1992) 100–111.
[27] L. Braun, et al., Differential response of nontumorigenic and tumorigenic human papillomavirus type 16-positive epithelial cells to transforming growth factor beta 1, Cancer Res. 50 (22) (1990) 7324–7332.
[28] C.D. Woodworth, et al., Transforming growth factor beta 1 supports autonomous growth of human papillomavirus-immortalized cervical keratinocytes under conditions promoting squamous differentiation, Cell Growth Differ. 7 (6) (1996) 811–820.

[29] J. Malejczyk, et al., Autocrine growth limitation of human papillomavirus type 16-harboring keratinocytes by constitutively released tumor necrosis factor-alpha, J. Immunol. 149 (8) (1992) 2702–2708.
[30] L.L.Villa, et al., Differential effect of tumor necrosis factor on proliferation of primary human keratinocytes and cell lines containing human papillomavirus types 16 and 18, Mol. Carcinog. 6 (1) (1992) 5–9.
[31] C.D. Woodworth, et al., Interleukin 1 alpha and tumor necrosis factor alpha stimulate autocrine amphiregulin expression and proliferation of human papillomavirus-immortalized and carcinoma-derived cervical epithelial cells, Proc. Natl. Acad. Sci. U. S. A. 92 (7) (1995) 2840–2844.
[32] H. zur Hausen, Papillomavirus infections—a major cause of human cancers, Biochim. Biophys. Acta 1288 (2) (1996) F55–F78.
[33] Cancer, I.I.I.C.o.H.a, India: human papillomavirus and related cancers, Fact Sheet 2018, 2019.
[34] C. de Martel, et al., Worldwide burden of cancer attributable to HPV by site, country and HPV type, Int. J. Cancer 141 (4) (2017) 664–670.
[35] R.L. Siegel, K.D. Miller, A. Jemal, Cancer statistics, 2019, CA Cancer J. Clin. 69 (1) (2019) 7–34.
[36] L.J. Rogers, M.A. Cuello, Cancer of the vulva, Int. J. Gynaecol. Obstet. 143 (Suppl. 2) (2018) 4–13.
[37] S. Rajaram, B. Gupta, Management of vulvar cancer, Rev. Recent Clin. Trials 10 (4) (2015) 282–288.
[38] G.J. Diorio, A.R. Giuliano, The role of human papilloma virus in penile carcinogenesis and Preneoplastic lesions: a potential target for vaccination and treatment strategies, Urol. Clin. North Am. 43 (4) (2016) 419–425.
[39] M.M. Symer, H.L. Yeo, Recent advances in the management of anal cancer, F1000Res 7 (2018).
[40] A.K. Ghosh, Cryosurgery of genital warts in cases in which podophyllin treatment failed or was contraindicated, Br. J. Vener. Dis. 53 (1) (1977) 49–53.
[41] A.K. Leung, et al., Penile warts: an update on their evaluation and management, Drugs Context 7 (2018), 212563.
[42] D.J. Wiley, et al., External genital warts: diagnosis, treatment, and prevention, Clin. Infect. Dis. 35 (Suppl. 2) (2002) S210–S224.
[43] G. Gross, Clinical aspects and therapy of anogenital warts and papillomavirus-associated lesions, Hautarzt 52 (1) (2001) 6–17.
[44] H.C. Fantasia, Sinecatechins ointment 15% for the treatment of external genital warts, Nurs. Womens Health 16 (5) (2012) 418–422.
[45] J. Doorbar, et al., Human papillomavirus molecular biology and disease association, Rev. Med. Virol. 25 (Suppl. 1) (2015) 2–23.
[46] C. Johansson, S. Schwartz, Regulation of human papillomavirus gene expression by splicing and polyadenylation, Nat. Rev. Microbiol. 11 (4) (2013) 239–251.
[47] J.T. Schiller, D.R. Lowy, Understanding and learning from the success of prophylactic human papillomavirus vaccines, Nat. Rev. Microbiol. 10 (10) (2012) 681–692.
[48] X.S. Chen, et al., Structure of small virus-like particles assembled from the L1 protein of human papillomavirus 16, Mol. Cell 5 (3) (2000) 557–567.
[49] I. Rubio, et al., The N-terminal region of the human papillomavirus L2 protein contains overlapping binding sites for neutralizing, cross-neutralizing and non-neutralizing antibodies, Virology 409 (2) (2011) 348–359.
[50] Y. Modis, B.L. Trus, S.C. Harrison, Atomic model of the papillomavirus capsid, EMBO J. 21 (18) (2002) 4754–4762.
[51] A.A. McBride, Replication and partitioning of papillomavirus genomes, Adv. Virus Res. 72 (2008) 155–205.

[52] J. Doorbar, The papillomavirus life cycle, J. Clin. Virol. 32 (Suppl. 1) (2005) S7–15.
[53] C.B. Buck, P.M. Day, B.L. Trus, The papillomavirus major capsid protein L1, Virology 445 (1–2) (2013) 169–174.
[54] P.M. Day, et al., The papillomavirus minor capsid protein, L2, induces localization of the major capsid protein, L1, and the viral transcription/replication protein, E2, to PML oncogenic domains, J. Virol. 72 (1) (1998) 142–150.
[55] M. Scheffner, et al., The E6 oncoprotein encoded by human papillomavirus types 16 and 18 promotes the degradation of p53, Cell 63 (6) (1990) 1129–1136.
[56] B.A. Werness, A.J. Levine, P.M. Howley, Association of human papillomavirus types 16 and 18 E6 proteins with p53, Science 248 (4951) (1990) 76–79.
[57] K. Uehara, et al., Co-expression of low-risk HPV E6/E7 and EBV LMP-1 leads to precancerous lesions by DNA damage, BMC Cancer 21 (1) (2021) 688.
[58] N. Dyson, et al., The human papilloma virus-16 E7 oncoprotein is able to bind to the retinoblastoma gene product, Science 243 (4893) (1989) 934–937.
[59] S. Cheng, et al., Differentiation-dependent up-regulation of the human papillomavirus E7 gene reactivates cellular DNA replication in suprabasal differentiated keratinocytes, Genes Dev. 9 (19) (1995) 2335–2349.
[60] F. Fehrmann, D.J. Klumpp, L.A. Laimins, Human papillomavirus type 31 E5 protein supports cell cycle progression and activates late viral functions upon epithelial differentiation, J. Virol. 77 (5) (2003) 2819–2831.
[61] M. Hummel, J.B. Hudson, L.A. Laimins, Differentiation-induced and constitutive transcription of human papillomavirus type 31b in cell lines containing viral episomes, J. Virol. 66 (10) (1992) 6070–6080.
[62] R. Dixit, C. Bhavsar, Y.S. Marfatia, Laboratory diagnosis of human papillomavirus virus infection in female genital tract, Indian J. Sex. Transm. Dis. AIDS 32 (1) (2011) 50–52.
[63] A.N. Burchell, et al., Chapter 6: epidemiology and transmission dynamics of genital HPV infection, Vaccine 24 (Suppl 3) (2006) S3/52–61.
[64] X. Castellsague, et al., Human papillomavirus (HPV) infection in pregnant women and mother-to-child transmission of genital HPV genotypes: a prospective study in Spain, BMC Infect. Dis. 9 (2009) 74.
[65] D. Tsakogiannis, et al., Molecular approaches for HPV genotyping and HPV-DNA physical status, Expert Rev. Mol. Med. 19 (2017), e1.
[66] J. Triplett, et al., Warts and all: Fingolimod and unusual HPV-associated lesions, Mult. Scler. 25 (11) (2019) 1547–1550.
[67] I.U. Park, C. Introcaso, E.F. Dunne, Human papillomavirus and genital warts: a review of the evidence for the 2015 Centers for Disease Control and Prevention sexually transmitted diseases treatment guidelines, Clin. Infect. Dis. 61 (Suppl 8) (2015) S849–S855.
[68] P. Khare, U. Sahu, Recent advances in the diagnostic methods of Leishmaniasis, in: M. Samant (Ed.), Pathogenesis, Treatment and Prevention of Leishmaniasis, ScienceDirect, 2021, pp. 45–62.
[69] D. Guenat, et al., Molecular diagnosis of human papillomaviruses (HPV): what test(s) in clinical practice? J. Gynecol. Obstet. Biol. Reprod. (Paris) 45 (9) (2016) 1009–1019.
[70] E.T.H. Fontham, et al., Cervical cancer screening for individuals at average risk: 2020 guideline update from the American Cancer Society, CA Cancer J. Clin. 70 (5) (2020) 321–346.
[71] Y. Ge, et al., HPV status in women with high-grade dysplasia on cervical biopsy and preceding negative HPV tests, J. Am. Soc. Cytopathol. 8 (3) (2019) 149–156.
[72] M. Schiffman, et al., Relative performance of HPV and cytology components of Cotesting in cervical screening, J. Natl. Cancer Inst. 110 (5) (2018) 501–508.
[73] A. Vives, M. Cosentino, J. Palou, The role of human papilloma virus test in men: first exhaustive review of literature, Actas Urol. Esp. (Engl. Ed.) 44 (2) (2020) 86–93.

[74] A.R. Giuliano, et al., Efficacy of quadrivalent HPV vaccine against HPV infection and disease in males, N. Engl. J. Med. 364 (5) (2011) 401–411.
[75] K.A. Szymonowicz, J. Chen, Biological and clinical aspects of HPV-related cancers, Cancer Biol. Med. 17 (4) (2020) 864–878.
[76] E.A. Mesri, M.A. Feitelson, K. Munger, Human viral oncogenesis: a cancer hallmarks analysis, Cell Host Microbe 15 (3) (2014) 266–282.
[77] B. Damania, A Virological perspective on Cancer, PLoS Pathog. 12 (2) (2016), e1005326.
[78] R.B.S. Roden, P.L. Stern, Opportunities and challenges for human papillomavirus vaccination in cancer, Nat. Rev. Cancer 18 (4) (2018) 240–254.
[79] K.P. Braaten, M.R. Laufer, Human papillomavirus (HPV), HPV-related disease, and the HPV vaccine, Rev. Obstet. Gynecol. 1 (1) (2008) 2–10.
[80] P. Bonanni, S. Boccalini, A. Bechini, Efficacy, duration of immunity and cross protection after HPV vaccination: a review of the evidence, Vaccine 27 (Suppl 1) (2009) A46–A53.
[81] M. Tommasino, The human papillomavirus family and its role in carcinogenesis, Semin. Cancer Biol. 26 (2014) 13–21.
[82] A.B. Moscicki, et al., Chapter 5: updating the natural history of HPV and anogenital cancer, Vaccine 24 (Suppl 3) (2006) S3/42–51.
[83] J.S. Smith, et al., Human papillomavirus type distribution in invasive cervical cancer and high-grade cervical lesions: a meta-analysis update, Int. J. Cancer 121 (3) (2007) 621–632.
[84] T.S. Adams, M.A. Cuello, Cancer of the vagina, Int. J. Gynaecol. Obstet. 143 (Suppl 2) (2018) 14–21.
[85] A.K. Sinno, et al., Human papillomavirus genotype prevalence in invasive vaginal cancer from a registry-based population, Obstet. Gynecol. 123 (4) (2014) 817–821.
[86] A. Gadducci, et al., Squamous cell carcinoma of the vagina: natural history, treatment modalities and prognostic factors, Crit. Rev. Oncol. Hematol. 93 (3) (2015) 211–224.
[87] M. Jit, et al., Cost-effectiveness of female human papillomavirus vaccination in 179 countries: a PRIME modelling study, Lancet Glob. Health 2 (7) (2014) e406–e414.
[88] L.C. Horn, et al., HPV-associated alterations of the vulva and vagina. Morphology and molecular pathology, Pathologe 32 (6) (2011) 467–475.
[89] H.Y. Ngan, et al., Abnormal expression and mutation of p53 in cervical cancer—a study at protein, RNA and DNA levels, Genitourin. Med. 73 (1) (1997) 54–58.
[90] T.H. Dellinger, et al., Surgical Management of Vulvar Cancer, J. Natl. Compr. Canc. Netw. 15 (1) (2017) 121–128.
[91] Y. Niu, et al., Clinical analysis of neoadjuvant chemotherapy in patients with advanced vulvar cancer: a STROBE-compliant article, Medicine (Baltimore) 97 (34) (2018), e11786.
[92] M.C. Bleeker, et al., Penile cancer: epidemiology, pathogenesis and prevention, World J. Urol. 27 (2) (2009) 141–150.
[93] J.S. Barnholtz-Sloan, et al., Incidence trends in primary malignant penile cancer, Urol. Oncol. 25 (5) (2007) 361–367.
[94] L.C. Kidd, et al., Relationship between human papillomavirus and penile cancer-implications for prevention and treatment, Transl. Androl. Urol. 6 (5) (2017) 791–802.
[95] L. Iorga, et al., Penile carcinoma and HPV infection (review), Exp. Ther. Med. 20 (1) (2020) 91–96.
[96] R. Assi, et al., Anorectal human papillomavirus: current concepts, Yale J. Biol. Med. 87 (4) (2014) 537–547.
[97] G. D'Souza, et al., Incidence and epidemiology of anal cancer in the multicenter AIDS cohort study, J. Acquir. Immune Defic. Syndr. 48 (4) (2008) 491–499.
[98] J.M. Palefsky, et al., HPV vaccine against anal HPV infection and anal intraepithelial neoplasia, N. Engl. J. Med. 365 (17) (2011) 1576–1585.

[99] H. Ram, et al., Oral cancer: risk factors and molecular pathogenesis, J. Maxillofac. Oral Surg. 10 (2) (2011) 132–137.
[100] E.L. You, M. Henry, A.G. Zeitouni, Human papillomavirus-associated oropharyngeal cancer: review of current evidence and management, Curr. Oncol. 26 (2) (2019) 119–123.
[101] S. Elrefaey, et al., HPV in oropharyngeal cancer: the basics to know in clinical practice, Acta Otorhinolaryngol. Ital. 34 (5) (2014) 299–309.
[102] S. Warnakulasuriya, Causes of oral cancer—an appraisal of controversies, Br. Dent. J. 207 (10) (2009) 471–475.
[103] N. Aggarwal, et al., Phytochemicals as potential Chemopreventive and chemotherapeutic agents for emerging human papillomavirus-driven head and neck cancer: current evidence and future prospects, Front. Pharmacol. 12 (2021), 699044.
[104] M.J. McCullough, C.S. Farah, The role of alcohol in oral carcinogenesis with particular reference to alcohol-containing mouthwashes, Aust. Dent. J. 53 (4) (2008) 302–305.
[105] G. D'Souza, et al., Oral sexual behaviors associated with prevalent oral human papillomavirus infection, J Infect Dis 199 (9) (2009) 1263–1269.
[106] A.R. Kreimer, et al., Oral human papillomavirus infection in adults is associated with sexual behavior and HIV serostatus, J Infect Dis 189 (4) (2004) 686–698.
[107] S.B. Chinn, J.N. Myers, Oral cavity carcinoma: current management, controversies, and future directions, J. Clin. Oncol. 33 (29) (2015) 3269–3276.
[108] E.I. Grussendorf-Conen, S. Jacobs, Efficacy of imiquimod 5% cream in the treatment of recalcitrant warts in children, Pediatr. Dermatol. 19 (3) (2002) 263–266.
[109] M.D. Greenberg, et al., A double-blind, randomized trial of 0.5% podofilox and placebo for the treatment of genital warts in women, Obstet. Gynecol. 77 (5) (1991) 735–739.
[110] E. Stockfleth, T. Meyer, The use of sinecatechins (polyphenon E) ointment for treatment of external genital warts, Expert Opin. Biol. Ther. 12 (6) (2012) 783–793.
[111] M. Brisson, et al., Population-level impact, herd immunity, and elimination after human papillomavirus vaccination: a systematic review and meta-analysis of predictions from transmission-dynamic models, Lancet Public Health 1 (1) (2016) e8–e17.
[112] A.C. Attia, J. Wolf, A.E. Nunez, On surmounting the barriers to HPV vaccination: we can do better, Ann. Med. 50 (3) (2018) 209–225.
[113] R. Sankaranarayanan, et al., Can a single dose of human papillomavirus (HPV) vaccine prevent cervical cancer? Early findings from an Indian study, Vaccine 36 (32 Pt A) (2018) 4783–4791.
[114] P.K. Braverman, HPV Vaccine in Adolescents, Pediatr. Ann. 48 (2) (2019) e71–e77.
[115] P.J. Maver, M. Poljak, Progress in prophylactic human papillomavirus (HPV) vaccination in 2016: a literature review, Vaccine 36 (36) (2018) 5416–5423.
[116] L. Zhai, E. Tumban, Gardasil-9: a global survey of projected efficacy, Antiviral Res. 130 (2016) 101–109.
[117] N.R. Chu, Therapeutic vaccination for the treatment of mucosotropic human papillomavirus-associated disease, Expert Opin. Biol. Ther. 3 (3) (2003) 477–486.
[118] B. Jabbar, et al., Antigenic peptide prediction from E6 and E7 Oncoproteins of HPV types 16 and 18 for therapeutic vaccine design using Immunoinformatics and MD simulation analysis, Front. Immunol. 9 (3000) (2018).
[119] A. Namvar, et al., In silico/In vivo analysis of high-risk papillomavirus L1 and L2 conserved sequences for development of cross-subtype prophylactic vaccine, Sci. Rep. 9 (1) (2019) 15225.
[120] A. Namvar, et al., Development of HPV(16,18,31,45) E5 and E7 peptides-based vaccines predicted by immunoinformatics tools, Biotechnol. Lett. 42 (3) (2020) 403–418.
[121] J. Wu, et al., Worldwide trend in human papillomavirus-attributable cancer incidence rates between 1990 and 2012 and Bayesian projection to 2030, Cancer 127 (17) (2021) 3172–3182.

[122] M. Lin, et al., Recent advances on the molecular mechanism of cervical carcinogenesis based on systems biology technologies, Comput. Struct. Biotechnol. J. 17 (2019) 241–250.
[123] S. Mikulicic, et al., Tetraspanin CD9 affects HPV16 infection by modulating ADAM17 activity and the ERK signalling pathway, Med. Microbiol. Immunol. 209 (4) (2020) 461–471.
[124] F.X. Bosch, et al., Prevalence of human papillomavirus in cervical cancer: a worldwide perspective. International biological study on cervical cancer (IBSCC) study group, J. Natl. Cancer Inst. 87 (11) (1995) 796–802.
[125] J. Ferlay, et al., Estimating the global cancer incidence and mortality in 2018: GLOBOCAN sources and methods, Int. J. Cancer 144 (8) (2019) 1941–1953.
[126] G. Curty, P.S. de Carvalho, M.A. Soares, The role of the Cervicovaginal microbiome on the genesis and as a biomarker of premalignant cervical intraepithelial neoplasia and invasive cervical Cancer, Int. J. Mol. Sci. 21 (1) (2019) 222.
[127] A. Ferenczy, et al., Human papillomavirus and HIV coinfection and the risk of neoplasias of the lower genital tract: a review of recent developments, Can. Med. Assoc. J. 169 (5) (2003) 431–434.

CHAPTER 2

Pathogenesis and immune response against HPV infection

Utkarsha Sahu and Prashant Khare
Department of Microbiology, All India Institute of Medical Sciences, Bhopal, Madhya Pradesh, India

1. Introduction

Human papillomavirus (HPV) is a quasi-enveloped double-stranded DNA (dsDNA) virus belonging to the *Papillomaviridae* family. HPVs are broadly divided into two categories, namely high-risk HPV (HR-HPV) types, including 16, 18, 31, 33, and 45, and low-risk (LR-HPV) types, including 6, 11, 40, 42, 43, 44, and 54 [1]. The HR-HPV strains are related to persistent viral infection causing skin lesions, genital warts, cervical intraepithelial neoplasia (CIN), and tumorigenesis, resulting in cervical cancer (CC). HPV types 16 and 18 are alone responsible for about 70% of cervical cancer cases worldwide [2]. HPV infection is predominant in sexually active people and mainly targets the epithelial cells. Mostly, the HPV infection is self-limiting; however, the failure of the host immune system to restrict HPV infection results in the incorporation of viral episomal DNA into the host chromosome leading to severe outcomes. The exact mechanism underlying the clearance of HPV infection by cellular immunity is still elusive. However, few HPV-induced lesions are not resolved by the host immunity and can persist for a long. Persistent HR-HPV infection might lead to malignancies [3]. The virus-infected host keratinocytes contain viral episomal DNA in the latent form. The multiplication, gene expression, and assembly of the viral genome are restricted to the dividing keratinocytes [3]. The different viral proteins facilitate the replication of the viral genome and evade immune recognition. The E1 protein behaves as a helicase and mediates the unwinding of viral DNA [4]. Later, the E6 and E7 oncoproteins are expressed that target the cell cycle regulatory factors. E6 interacts with p53 and promotes its degradation [5,6]. Similarly, E7 interacts with retinoblastoma (Rb) and releases the E2F transcription factor, thereby keeping the cell cycle in the S phase [5,7,8]. The most immunogenic late proteins

Immunopathology, Diagnosis, and Treatment of HPV-induced Malignancies
https://doi.org/10.1016/B978-0-323-91797-1.00006-5

(L1 and L2) are synthesized at last in order to delay immune recognition. The major and minor capsid proteins L1 and L2 are responsible for viral packaging into mature virus particles. Besides targeting the cell cycle genes, HPV also downregulates toll-like receptor 9 (TLR9) expression, major histocompatibility complex I (MHC I), and the interferon response. Once the viral episome is inserted into the host chromosomes, it leads to oncogenic progression leading to CIN and cancer. The exact mechanism responsible for the failure of cell-mediated immunity to nullify HPV infection is not fully explored. Hence, understanding the mechanism of pathogenesis, including viral attachment, entry, and replication along with evasion of the immune response, would be beneficial to identify novel drug targets and potential therapeutic vaccine candidates against HPV.

2. Pathogenesis and immune response associated with HPV infection

2.1 HPV structure, replication, and pathogenesis

The 8 kb viral genome of HPV encodes six early (E1, E2, E4, E5, E6, and E7) genes and two late (L1 and L2) genes and possesses noncoding long control region (LCR) [9,10] (Fig. 1).

The early proteins are associated with viral pathogenicity, whereas the late proteins form the viral capsid [9,10]. The genomic structure of HPV is covered by a capsid of 72 pentameric capsomeres formed by viral-coded 55 kDa L1 protein and 70 kDa L2 protein. The L1 protein alone or along

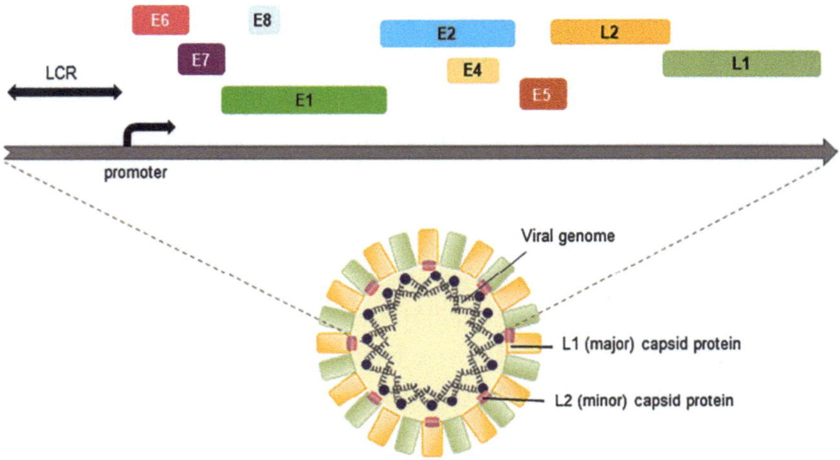

Fig. 1 Structure of human papillomavirus genome.

with the L2 protein can form HPV virus-like particles (VLPs) in non-mammalian or mammalian expression systems. The assembled HPV virion has 52–55 nm of diameter, sedimentation coefficient value of 300, and 1.34 g/mL of density [11].

The entry of HPV into the host epithelium is mediated either via microabrasions [12] or by targeting the single-layered cellular junction between ectocervix and endocervix [13]. Once the viruses enter, an effective infection is established only if they infect the dividing stem or basal epithelial cells, which are mainly reported for HR-HPV strains [14]. The initial binding of HPV is through the heparin sulfate proteoglycans (HSPGs) [15]. After the binding of HPV with HSPG, the viral capsid undergoes a conformational change and the N terminal of L2 is exposed at the surface of the virus [16]. This N terminus is later cleaved by PC5/6 and/or furin that enables binding with a secondary receptor on the surface of the plasma membrane [17,18]. In addition, other studies have also reported the involvement of integrins (α6 integrin) [19–21], epidermal growth factor (EGF) receptors (EGFRs) [19], laminins [22], tetraspanin-enriched membrane microdomains [23], vimentin [24], and the annexin-A2 heterotetramer [25,26]. The choice of receptor depends on the type of target cells, HPV genotype, or combination of different receptors during infection [27]. HPV virions are internalized into the target cell via endocytosis that shares a resemblance with micropinocytosis [28]. The virus moves through membrane-bound constituents of the cytoplasm and trans-GC and ER networks [28–30]. At last, the viral episome is transported into the host cell nucleus by tubulin-mediated pathway through nuclear pores or breaking the nuclear membrane in mitosis [14,31]. HPV takes around 24 h to reach the nucleus after the attachment to the target cell, followed by the attachment of L2 and viral genomes to promyelocytic leukemia (PML) nuclear bodies [32]. The PMLs are utilized by the DNA viruses to carry transcription. Many viruses destroy the PMLs after infection; however, HPVs retain PML integrity to develop nuclear infection [33,34].

The HPV replication cycle can be distributed into three different stages, namely establishment, maintenance, and amplification [35] (Fig. 2).

As soon as infection is established, the viral genome translocates into the nucleus of the host cell, and then, limited amplification of viral DNA occurs to initiate the infection. Later, the replication of the viral genome occurs in the basal keratinocytes undergoing active mitosis to maintain a constant copy number of the viral genome. After the differentiation of mature keratinocytes, the rate of replication increases to upregulate the copy number of

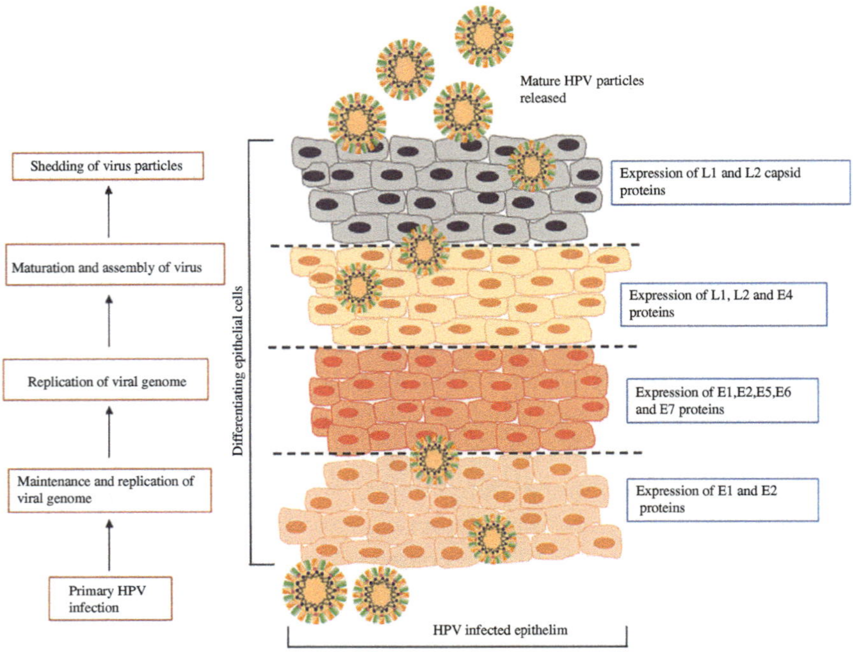

Fig. 2 HPV life cycle in human epithelial cells.

viral DNA, followed by their assembly as mature viral particles [35]. HPV genome has eight different open reading frames (ORFs) comprising three functional parts, namely the early region coding for E1-E7 proteins, the late region coding for L1 and L2 proteins, and a cis-acting LCR region crucial for DNA replication and transcription [35] (Fig. 1). The E region codes for genes necessary for viral DNA replication, evasion of host immunity, and tumorigenesis [36]. The origin of replication for HPV comprises E1 and E2 binding sites and E1 and E2 proteins. E1 is an ATP-dependent helicase that unwinds the origin of replication and provides access to cellular replication proteins [37]. The E2 loading protein forms a dimer with E1 and binds at its respective site. After loading, the E1–E2 dimer dissociates and E1 forms a helicase with a double hexameric ring encircling the DNA [38,39]. The early expression of E2 aids in regulating the viral E6–E7 proteins enabling the infected cells' survival [5]. The initial cycle of replication produces 50–100 copies of viral episome per nucleus [40]. This limited increase of the viral episome is regulated by the E8∧E2 protein through NCoR/SMRT complex [41]. These steps are part of genome replication and maintenance.

The lower level of viral protein expression occurs at basal cells to avoid stimulating the immune system [42,43]. This is carried out by transcriptional repression of the P97 promoter by E2 via avoiding the binding of transcription factors and changing the chromatin conformation [44,45]. Cell division of basal epithelial cells infected with HPV results in cells capable of differentiating and migrating toward the upper layers of the epithelium [46]. These epithelial cells contain viral genes as they migrate upward by taking viral genes with them. HPV has developed a certain mechanism to replicate its genome parallel to epithelial differentiation [47]. HPVs have an early promoter responsible for early protein expression of E6 and E7 viral proteins [48,49], crucial for the normal viral replication cycle. The E6 viral protein is necessary for maintaining the episomal genome [50,51]. E6 can further nullify the p53 activity by binding and recruiting CBP/p300 histone acetyltransferase [52,53] and also seize p53 into the cell cytoplasm and prevent it from entering the nucleus [54]. E7 expression triggers G1/S phase checkpoints in the keratinocytes undergoing differentiation, thus enabling continuous viral replication in the infected keratinocytes [49]. E7 initiates the cell cycle by degrading or releasing Rb and different pocket proteins like p107 and p130 from the E2F complex. This results in free transcription factor E2F that activates different cell cycle factors such as cyclins E and A, thereby initiating the transition of the G phase toward the S phase [5,7,8]. HPV E7 protein can act together with the E2F inhibitory complex, resulting in its inhibition and finally promoting the growth-associated genes [55]. E7 also has different inhibitory and stimulatory effects on various transcription factors like NF-κB, signal transducer and activator of transcription 1 (STAT1), E2F transcription factor 6 (E2F6), interferon regulatory factor 1 (IRF1), E2F transcription factor 1 (E2F1), TATA-binding protein (TBP), c-Fos, SMAD2/3, Miz 1, c-Jun, c-Myc, and B-Myb [56]. Altogether, this suggests the crucial role of E7 in modulating the transcription of the cell. Nevertheless, the E7 expression in different layers of keratinocytes and the level of expression influence these activities. Also, the expression of E7 is always along with E6 because of the bicistronic property of their coding region [57]. This means that the activity of one would influence the other as they are often reported to act together mainly in evasion of host immune response [43]. Habitually, the cells answer back to spontaneous proliferation by apoptosis. To overcome this, the HR-HPV strains express E6 oncoprotein that binds to E6-associated protein (E6AP), p53, and ubiquitin ligase and marks it for degradation [5,6]. The other early proteins E4 and E5 perform a supplementary role in viral replication and life cycle. E4 regulates the processing of RNA in the infected cell via SR protein kinase 1 [58], and it can also bind with DEAD-box RNA helicase that might aid the posttranscriptional regulation of gene expression [59]. One of the most vital functions of E4 is restructuring of cytokeratin filaments making the cells fragile and aiding the release of virions [60–62]. E5, on the other hand, helps in immune evasion by downregulation of antigen presentation by MHC I [63]. E5 also affects different signaling pathways like epidermal growth factor receptor (EGFR), extracellular signal-regulated kinase (ERK), map kinase (MAPK), p38, and AKT pathways [64].

The DNA damage response pathway is a vital step for the completion of HPV replication and expression of the late genes. The major sensors of damaged DNA are two kinases—ataxia telangiectasia and Rad3-related (ATR) and ataxia telangiectasia-mutated (ATM) kinase that also regulate the DNA repair mechanism by recruiting different replication and repair factors [65,66]. During the early stage of replication, ATM is essential to establish and maintain the HPV genome in the basal epithelial cells [65,66]. During the late phase, the E7 oncoprotein activates the ATM-mediated DNA damage pathway for viral gene amplification in the actively differentiating keratinocytes [67]. The viral E1 protein activates a DNA damage response resulting in the recruitment of ATM/ATR at the viral replication sites [68–70]. The exact mechanism underlying the HPV-regulated ATR mechanism is still elusive. However, E7 may induce cell cycle in the differentiating epithelial cells by indirectly activating the ATR via the replication stress mechanism [66]. Apart from the ATM/ATR pathway, the viral early proteins E1 and E2 form viral replication factories comprising several proteins involved in DNA damage repair [70].

The viral circular DNA is replicated along with the cellular DNA and is equally divided into the newly formed daughter cells by hitching the viral genome into host chromosomes [71]. Brd4 is the most explored anchor for viral genomes to host chromosomes [72]. However, the other alternatives for the attachment of HPV genomes include MKlp2 [73], TopBP1 [42], and ChlR1 [74]. Contrariwise, the E1–E2 complex by itself can be enough for few HPVs to localize with the cellular chromosomes [71].

After the replication of viral DNA, the viral capsid protein synthesis occurs in the granular epithelial layer. The synthesis of very immunogenic capsid protein is initially at a halt to avoid immune recognition [75]. The capsid protein is expressed from the late promoters (P811 and P670) in cells promoting the replication of the HPV genome [76]. Therefore, inhibition of expression of the capsid protein is not carried out at the transcription stage. Different posttranscriptional mechanisms such as polyadenylation, alternative splicing, translation control, and mRNA stability are related to the delayed expression of capsid protein [76,77]. Different RNA-binding proteins like splicing factor 1 (SRSF1 (SF2/ASF)) [78], mRNA stability regulator HuR [79], U1 small nuclear ribonucleoprotein (U1snRNP) [80], and hnRNP A1 [81] bind and regulate the expression of capsid protein. The minor capsid protein L2 is translated before L1 and is translocated inside the nucleus [82]. In comparison, L1 is self-assembled in the cytoplasm to form pentameric capsomeres and is later transported into the nucleus. After the synthesis, the viral genome translocates to the PML via E2, which is bound with the LCR region and enables them to be near the capsid protein that is ready for assembly [32]. Now, L1 and L2 proteins interact with each other and the assembly of virions takes place. L2 is mainly related to the assembly of virions at the capsomere center [83,84]. Newly formed virus particles are released from the dead squamous cells shedding off from the epithelial layer. These virions are viable sometimes and infect the nearby cells [61] (Fig. 1).

2.2 Immune response against HPV

2.2.1 Innate immune response against HPV

Innate immunity is the first line of protection against invading pathogens. It is nonspecific and short-lasting. Epithelial cells were initially thought to work as a mechanical barrier. Recently, a diverse role of epithelial cells in host immunity has been explored. Keratinocytes are a type of epithelial cells that form the superficial layer of the skin [85]. Keratinocytes can produce lower concentrations of different chemokines, growth factors, and pro-inflammatory cytokines upon different stimulations [86,87]. The major antiproliferative and antiviral cytokines involved in HPV infection include tumor necrosis factor (TNF), type I interferons (IFN-α and β), and transforming growth factor β (TGF-β) [88]. The studies performed on the inhibitory and stimulatory activities of cytokines were mostly in vitro and depend on the experiment conditions, which could explain the mixed results [88].

TGF-β is one such example of these mixed findings. In HPV16 and HPV18 immortalized cells, TGF-β was reported to be inhibited as a result of HPV infection [89,90]. The inhibition of cell proliferation in HPV16 infection occurs at a transcription level. The outcomes of exogenous TGF-β could be obstructed with cycloheximide, signifying a need for negative regulatory protein synthesis [89,91]. Studies performed by Woodworth et al. highlighted an autocrine regulation role of TGF-β in regulating HPV gene expression [91]. In addition, TGF-β was also reported to limit HPV proliferation by reducing the RNA level of c-*myc* gene that is crucial for cell proliferation [90]. Along with *c-myc*, TGF-β also regulated other genes associated with cell proliferation. In HPV11 xenografts, the upregulation of TGF-β was directly correlated with the downregulation of different genes such as c-*jun*, c-Ha-*ras*, *bcl-2* and c-*myc* and nuclear factor-κB (NF-κB), implying a negative relation between TGF-β upregulation and inhibition of HPV [92]. Furthermore, TGF-β treatment of HPV16-transformed cells downregulated NF-κB and *bcl2* expression [93]. In another study, TGF-β stimulated the growth of HPV-transformed cells grown in media stimulating the differentiation of squamous cells. This growth is observed only in HPV-induced transformed cells, not in normal uninfected keratinocytes [94].

TNF, another cytokine secreted by keratinocytes, has both stimulatory and antiproliferative effects on HPV-infected cells. In HPV16-infected epithelial cells, TNF serves as an antiproliferative agent; however, this was not observed in HPV18-infected cells [95,96]. The antiproliferative effect was due to cell cycle arrest at the G0/G1 stage [97]. Similar to TGF-β, TNF also suppresses the expression of HPV16 E6 and E7 at the transcription level in the human keratinocyte cell line [98]. In addition, TNF also has an autocrine loop mediated by amphiregulin in some HPV16- and HPV18-infected CC cell lines, suggesting the possibility of escaping growth inhibition [99].

IFN- α, β, γ are already investigated for their antiproliferative property [100,101]. Further investigations in HPV-infected cells revealed that IFN-α inhibited HPV16-transformed keratinocytes at 10- to 100-fold lower concentrations as compared to that needed for inhibition of normal keratinocytes [100]. IFN-α also restricts the expression of HPV16 E7; however, it does not inhibit the transcription of E7 and translation of E6 protein. This implies that the growth inhibition by IFN-α is through E7 protein [100]. Other studies have reported that IFN-α and IFN-γ limit transcription of HPV18 E6 and E7 oncoproteins in the HeLa cell line [102,103]. The fluctuating effects of IFNs depend on different experimental variables, which are virus- and cell line-specific. In addition, one study reported that apart from growth inhibition, IFN-γ not IFN-α restricts the transcription of viral E6 and E7 genes in HPV16-, HPV18-, and HPV33-infected keratinocytes [104]. In contrast, another study showed that IFN-β downregulated E6 and E7 transcription in the HPK-1A cell line, while IFN-α and IFN-γ did not affect these cells [105]. Moreover, IFN-β also had cytopathic effects on the HPK-1A cell line and IFN-α did not, even though they share a common receptor [105,106]. This observation implies the ability of cell lines to differentiate between two cytokines sharing common receptors and also the response of different cytokines varies in different cell lines [105]. Altogether, various studies support the idea that various cytokines such as TGF-β, IFN α, β, γ, IL-1, and TNF play a crucial role in monitoring the growth in HPV-infected epithelial cells and might be associated with the persistence of viral infection, disease severity, and transformation [88].

2.2.2 Adaptive immune response against HPV

The different cellular components related to the adaptive immune response against HPV infections are well-documented [107–109]. These primarily include Langerhans cells, clonally expanded T cells, and other accessory cells. The Langerhans cells are reported to be linked with different cutaneous manifestations causing viral infections such as HPV. The major role of Langerhans cells is to capture and present the antigens to naïve T cells [88]. An abridged Langerhans count has been observed in lesions caused due to HPV infection [108]. A study reported significant reductions in $CD1^+$ Langerhans in the hand warts and cutaneous plantar normal epithelium; however, the Langerhans cells were unaffected in the laryngeal and mucosal genital condylomas [110]. Instead, other studies have reported reduced Langerhans cells in condylomas and genital HPV infection [109,111,112]. The decrease in the population of Langerhans cells in HPV-infected tissues might be due to the relocation of Langerhans cells to lymph nodes for presenting antigens to the naïve T cells [108].

Once the antigens are seized by Langerhans cells, their recognition still continues till the migration of cells to lymph nodes. Cytokines secreted by the keratinocytes and Langerhans cells promote this process [87]. The major cytokines promoting the movement of Langerhans cells include TNF and IL-1α from keratinocytes and IL-1β secreted by Langerhans cells [87]. Other cytokines secreted by keratinocytes such as granulocyte-macrophage colony-stimulating factor (GM-CSF) stimulate maturing of Langerhans cells to dendritic cells [113].

After reaching the dripping lymph nodes, the Langerhans cells display antigens to naïve T cells. The role of T lymphocytes in protection against HPV-induced lesions is well-established [114–116]. HPV16 is the most prevalent oncogenic strain causing CC. Thus, most studies regarding T cell response counter to HPV infection are performed on HPV16 [117]. E6 and E7 oncoproteins are widely studied. In few cross-sectional studies, the response against these antigens was more prevalent in normal subjects as compared to patients with squamous intraepithelial lesions (SIL) [116,118].

However, this response is still debated as other studies reported opposite results [114,119]. Other studies also examined the influence of HPV16 E6 and E7 oncoproteins on T cell proliferation. The longitudinal studies had mixed results. The first study proposed that patients displaying a positive response to E6 and E7 peptides cleared SIL and HPV infection [120]. On the contrary, the second study reported that patients with progressing disease and SIL displayed T cell proliferation and IL-2 secretion [115,121].

3. Evasion of host immune response

HPV uses passive and aggressive immune evasion techniques to elope the host immunity. The passive technique used by HPV involves downregulation of antigen production during the early stage of the life cycle. During this phase, the viral protein expression is minimum and is quickly transported to the nucleus, thereby minimizing the antigen presentation [122]. However, during the later phase, the capsid protein expression increases that is highly immunogenic. The capsid proteins quickly shed off from the outer epithelium having low density of APCs, thereby escaping the immune system [123]. The aggressive strategies of immune evasion are carried out by the viral oncoproteins. They possess a high binding affinity with different immune regulators and inhibit their expression and signaling pathways in the infected keratinocytes. In addition, the viral infection triggers a pro-inflammatory environment that aids viral persistence [124,125]. The infected cells are unable to signal the neighboring immune cells, causing immunosuppression that facilitates cancer progression [126].

Keratinocytes express different pathogen recognition receptors (PRRs) that identify different pathogen-associated molecular patterns of the virus [11]. Toll-like receptor 9 (TLR9), a PRR, recognizes the viral dsDNA and initiates the downstream antiviral signaling [127]. The viral oncoproteins employ different techniques to downregulate TLR9 expression. The E7 protein of HPV38 causes recruitment of EZH2 histone-modifying enzyme at TLR9 promoter, thereby causing histone methylation and inhibiting TLR9 expression [128]. Likewise, E7 of HPV16 reduces TLR9 expression by recruiting histone demethylase JARID1B and histone deacetylase HDAC1 at the TLR9 promoter [129].

Apart from TLR9, the viral DNA is also recognized by cyclic guanosine monophosphate (GMP)-adenosine monophosphate (AMP) synthase (cGAS). After recognition by cGAS, downstream signaling is initiated via STING adapter protein. The E7 protein of HPV18 restricts STING through the LXCXE motif, thereby downregulating the pro-inflammatory cytokines in keratinocytes [130]. The direct or inverse correlation of HPV susceptibility with TLR9 expression is still debatable. For instance, the high TLR9 expression in the cervical cells showed reduced susceptibility toward HPV infection as compared to low TLR9-producing cells [131]. In contrast, increased TLR9 expressions are reported in different studies performed on cervical cancer patients [132,133].

IFN response is the major antiviral mechanism responsible for effective viral clearance. HPV has evolved different mechanisms to nullify the functioning of IFNs [134,135]. The PRR signals activate the TNF-receptor-associated factor (TRAF), followed by the phosphorylation of different interferon regulatory transcription factors (IRFs). These phosphorylated IRFs form dimers and translocate inside the nucleus and initiate IFN production [136]. The HPV16 E6 protein binds to IRF and prevents it to translocate into the nucleus, thereby inhibiting type I IFN response against HPV [134]. In addition, the E7 protein binds with IRF1, recruits HDAC at IFN-β promoter, and prevents transcription of IFN-β [135].

HPV oncoproteins can potentially bind to tyrosine kinase 2 (TYK2) to avoid phosphorylation activation of STAT1 and STAT2, which are crucial transcription factors required for type I IFN response [137]. Furthermore, the HPV E6 protein also impairs the transcription and translation of STAT1 [138,139], while E7 prevents binding of IRF9 to phosphorylated STAT1-STAT2 dimer, thus preventing the translocation of the complex into the nucleus [140,141]. Apart from targeting the IFN mechanism, HPV also up-regulated ubiquitin C-terminal hydrolase L1 (UCHL1), a deubiquitinating enzyme that targets the polyubiquitin chains of TRAF3 linked at the K63 position and deactivates it [142]. Downregulation of the NF-κB pathway is another approach employed by the HPV to nullify host immunity [143]. E6 and E7 oncoproteins from HPV16-infected keratinocytes displayed NF-kB inhibition [143]. In addition, these proteins can bind to NF-kB coactivators and downregulate the NF-kB signaling [52,53,144]. HPV infection also triggers MHC I downregulation. The E7 protein from HPV interacts with the RXR-beta binding region at the MHC I promoter, recruiting HDACs and blocking the expression of MHC I [145,146]. A similar mechanism is employed by E7 to block the different components of antigen processing machinery (TAP1 and LMP2) [145,147]. Unlike E7, the E5 protein from HPV decreases MHC I expression by inhibiting the MHC I transportation toward the cell surface by directing the endoplasmic reticulum and Golgi (ER-GC) complex [148,149].

HPV impairs the host adaptive immune response by targeting the cytotoxic T lymphocytes (CTLs) and regulatory T cells (Tregs). Naive cytotoxic T cells ($CD8^+$ T cells) interact with mature antigen-presenting cells (APCs) and differentiate into CTLs. These CTLs migrate at the site of infection and kill the infected cells. Keratinocytes infected with HPV have reduced surface MHC I expression and thereby avoid CTL recognition [150,151]. During prolonged viral infection, the recruitment of Tregs can aid immune evasion and pathogenesis. Studies performed on Tregs from tumors and blood of persistently HPV-infected patients revealed higher $CD4^+$ $CD25^+$ $Foxp3^+$ Tregs as compared to recovered patients [152,153]. In addition, few studies have observed the presence of Tregs specific for E6 and E7 oncoproteins in cervix and lymph nodes [154,155]. As a result of in vitro stimulation with HPV antigen, the HPV-specific Tregs suppressed the proliferation of T cells and cytokine production, thereby creating an immunosuppressed state [154,155]. However, the precise mechanism underlying HPV-mediated activation and proliferation of Tregs is unexplored. Different factors present in the local immune setting might act together to aid Treg proliferation and activation [126]. The phenomenon of $CD8^+$ T and $CD4^+$ T cell infiltration is observed in different HPV-associated tumors; however, it is not efficient for the elimination of HPV-infected cells. The following mechanism for this inefficiency remains unexplored. Few studies have reported the transition of Th1 to Th2 immune response in persistent HPV-associated lesions. Additionally, patients suffering from higher-grade cervical lesions reported increased Th17 and Treg responses as compared to patients with repressed lesions [126]. Together, these tempered immune responses may contribute to reducing CTL response and subsequently in evasion of immune response of HPV in the tumor microenvironment. Overall, the infiltration of $CD8^+$ and $CD4^+$ T cells was observed in HPV-induced tumors, yet they are inefficient in eradicating HPV-infected cells.

4. Conclusion

HPV infection and its correlation with different malignancies have been the topic of research for past 50 years. Our understanding of HPV pathogenesis and evasion of host immunity is increasing rapidly. There are a series of events in the HPV pathogenesis and host immunity against HPV infection; however, a detailed understanding is elusive. A molecular insight of HPV pathogenesis is needed for developing effective therapeutics to prevent and/or treat HPV infection [1]. Different viral proteins are responsible for attachment, entry, replication, and immune evasion of HPVs. The attachment and entry of HPV in host keratinocytes are mediated by L1 and L2 capsid proteins [156–158]. Later, the virus replication and host immune evasion are carried out by the E proteins. E1 and E2 proteins are crucial for initiating viral replication [4,39,41,68,70,71,159]. This is followed by E6 and E7 oncoprotein expression. The HPV E6 protein binds with the p53 tumor suppressor protein and recruits different ubiquitination enzymes for p53 degradation to prevent cell death [160]. The viral E7 protein interacts with the Rb protein and releases free E2F transcription factor and S-phase-promoting factors [55]. The oncoproteins E6 and E7 are key factors associated with immune evasion and tumor establishment and progression [161]. Restriction of the viral genome cycle during the infection cycle by the E2 protein can be helpful in downregulating E6 and E7 expression. The role of other E proteins such as E4 and E5 is less explored. The HPV E4 protein is responsible for making the infected cells fragile for easy release of mature virus particles [61,62]. The HPV16 E5 protein forms a viroporin that possesses the activity of ion channel, which may be associated with E5's augmentation of EGFR activity in the infected cells. However, further studies are required to clarify this role of E5 [162]. The keratinocytes can produce lower concentrations of different chemokines, growth factors, and pro-inflammatory cytokines after stimulations [86,87]. The major antiproliferative and antiviral cytokines produced in HPV infection include TGF-β, TNF, and IFN-α and IFN-β [87]. Majority of these studies were carried out in cell lines whose outcomes may vary with the maintaining environment and infected strains. Thus, the results of such studies are confusing. For instance, TGF-β was reported to show both pro- and antiproliferative functions during HPV infection [90,92]. Similar findings were reported by the studies done on TNF response in HPV infection [95,96,99]. IFNs are the major antiviral cytokines that limit viral infections. In HPV infection, the IFN response was significantly downregulated [102,103,105,106]. The cellular immune response against HPV is mainly carried out by Langerhans cells, clonally expanded T cells, and other accessory cells [105–107]. The main role of Langerhans cells is to capture and present the antigens to naïve T cells [88]. Once the antigens are seized by Langerhans cells, their recognition continues until the migration of cells to lymph nodes. Cytokines produced by the keratinocytes (TNF and IL-1α) and Langerhans (IL-1β) cells promote this process [86]. The presentation of antigens to the naïve T cells leads to their differentiation. The proliferation of T cells in response to HPV antigens is debated [115,121,163,164]. However, so far, the specific role of T helper cells in triggering antiviral response during HPV infection is unclear. They contribute in the development of CTL and antibody production by B cells [165–168]. Conclusively, virus deploys different techniques to evade host immune response.

The HPV early proteins play a crucial role in this evasion. The virus expresses its highly immunogenic capsid protein at last to avoid immune recognition [122]. Also, E6 and E7 oncoproteins halt the cell cycle in the S phase by interacting with P53 and Rb protein. Also, HPV downregulates TLR9 expression by recruiting histone-modifying enzymes at its promoter [128,129]. Secondly, it targets STING adapter proteins to downregulate cGAS-mediated signaling [130]. IFN is the major antiviral cytokine responsible for viral clearance. The HPV16 E6 protein binds with the IRF and prevents it from translocating the nucleus and thereby restricting an effective IFN response [134]. In addition, the E7 protein binds with IRF1, recruits HDAC at IFN-β promoter, and prevents transcription of IFN-β [135]. In addition, HPV oncoproteins also bind with STAT1 and STAT2 and prevent their translocation into the nucleus and thus nullify an effective IFN response [138,139]. In addition, these proteins are able to bind with NF-kB coactivators and downregulate the NF-kB signaling [52,53,144] to prevent downstream cytokine response. The viral proteins also moderate the expression of MHC I by different pathways to evade the host immunity [148,149].

The fine-tuning between the early proteins with cell cycle and apoptosis to aid the amplification of the viral genome and avoiding initial integration into the host genome is important for completing the viral infection cycle. Still, we lack an understanding of different aspects of HPV pathogenesis. Also, most of the studies have focused on the replication and immune evasion of HR-HPV strains, while the LR-HPV strains are not fully explored. Finally, further research is needed to comprehend the cross talk between HPV pathogenesis and host immune response for developing more satisfactory preventive and therapeutic measures.

References

[1] M.S. Longworth, L.A. Laimins, Pathogenesis of human papillomaviruses in differentiating epithelia, Microbiol. Mol. Biol. Rev. 68 (2) (2004) 362–372.

[2] G.M. Clifford, et al., Comparison of HPV type distribution in high-grade cervical lesions and cervical cancer: a meta-analysis, Br. J. Cancer 89 (1) (2003) 101–105.

[3] H. zur Hausen, Papillomavirus infections—a major cause of human cancers, Biochim. Biophys. Acta 1288 (2) (1996) F55–F78.

[4] M.G. Frattini, L.A. Laimins, Binding of the human papillomavirus E1 origin-recognition protein is regulated through complex formation with the E2 enhancer-binding protein, Proc. Natl. Acad. Sci. U. S. A. 91 (26) (1994) 12398–12402.

[5] C.A. Moody, L.A. Laimins, Human papillomavirus oncoproteins: pathways to transformation, Nat. Rev. Cancer 10 (8) (2010) 550–560.
[6] D. Martinez-Zapien, et al., Structure of the E6/E6AP/p53 complex required for HPV-mediated degradation of p53, Nature 529 (7587) (2016) 541–545.
[7] D.L. Jones, R.M. Alani, K. Münger, The human papillomavirus E7 oncoprotein can uncouple cellular differentiation and proliferation in human keratinocytes by abrogating p21Cip1-mediated inhibition of cdk2, Genes Dev. 11 (16) (1997) 2101–2111.
[8] K. Huh, et al., Human papillomavirus type 16 E7 oncoprotein associates with the cullin 2 ubiquitin ligase complex, which contributes to degradation of the retinoblastoma tumor suppressor, J.Virol. 81 (18) (2007) 9737–9747.
[9] J. Doorbar, et al., Human papillomavirus molecular biology and disease association, Rev. Med.Virol. 25 (S1) (2015) 2–23.
[10] J. Doorbar, Papillomavirus life cycle organization and biomarker selection, Dis. Markers 23 (4) (2007) 297–313.
[11] R. Kirnbauer, et al., Papillomavirus L1 major capsid protein self-assembles into virus-like particles that are highly immunogenic, Proc. Natl.Acad. Sci. U. S. A. 89 (24) (1992) 12180–12184.
[12] J. Doorbar, The papillomavirus life cycle, J. Clin.Virol. 32 (Suppl 1) (2005) S7–15.
[13] M. Herfs, et al., A discrete population of squamocolumnar junction cells implicated in the pathogenesis of cervical cancer, Proc. Natl. Acad. Sci. U. S. A. 109 (26) (2012) 10516–10521.
[14] D. Pyeon, et al., Establishment of human papillomavirus infection requires cell cycle progression, PLoS Pathog. 5 (2) (2009), e1000318.
[15] J.T. Schiller, P.M. Day, R.C. Kines, Current understanding of the mechanism of HPV infection, Gynecol. Oncol. 118 (1 Suppl) (2010) S12–S17.
[16] M. Bienkowska-Haba, H.D. Patel, M. Sapp, Target cell cyclophilins facilitate human papillomavirus type 16 infection, PLoS Pathog. 5 (7) (2009), e1000524.
[17] R.M. Richards, et al., Cleavage of the papillomavirus minor capsid protein, L2, at a furin consensus site is necessary for infection, Proc. Natl. Acad. Sci. 103 (5) (2006) 1522–1527.
[18] R.C. Kines, et al., The initial steps leading to papillomavirus infection occur on the basement membrane prior to cell surface binding, Proc. Natl. Acad. Sci. U. S. A. 106 (48) (2009) 20458–20463.
[19] Z. Surviladze, A. Dziduszko, M.A. Ozbun, Essential roles for soluble virion-associated heparan sulfonated proteoglycans and growth factors in human papillomavirus infections, PLoS Pathog. 8 (2) (2012), e1002519.
[20] C.Y. Abban, P.I. Meneses, Usage of heparan sulfate, integrins, and FAK in HPV16 infection, Virology 403 (1) (2010) 1–16.
[21] M. Evander, et al., Identification of the alpha6 integrin as a candidate receptor for papillomaviruses, J.Virol. 71 (3) (1997) 2449–2456.
[22] T.D. Culp, et al., Keratinocyte-secreted laminin 5 can function as a transient receptor for human papillomaviruses by binding virions and transferring them to adjacent cells, J.Virol. 80 (18) (2006) 8940–8950.
[23] K.D. Scheffer, et al., Tetraspanin CD151 mediates papillomavirus type 16 endocytosis, J.Virol. 87 (6) (2013) 3435–3446.
[24] G. Schäfer, et al., Vimentin modulates infectious internalization of human papillomavirus 16 Pseudovirions, J.Virol. 91 (16) (2017).
[25] A.W. Woodham, et al., The S100A10 subunit of the annexin A2 heterotetramer facilitates L2-mediated human papillomavirus infection, PLoS One 7 (8) (2012), e43519.
[26] A. Dziduszko, M.A. Ozbun, Annexin A2 and S100A10 regulate human papillomavirus type 16 entry and intracellular trafficking in human keratinocytes, J.Virol. 87 (13) (2013) 7502–7515.

[27] S.V. Graham, The human papillomavirus replication cycle, and its links to cancer progression: a comprehensive review, Clin. Sci. (Lond.) 131 (17) (2017) 2201–2221.
[28] S. DiGiuseppe, et al., Cruising the cellular highways: how human papillomavirus travels from the surface to the nucleus, Virus Res. 231 (2017) 1–9.
[29] P.M. Day, et al., Identification of a role for the trans-Golgi network in human papillomavirus 16 pseudovirus infection, J. Virol. 87 (7) (2013) 3862–3870.
[30] W. Zhang, et al., Vesicular trafficking of incoming human papillomavirus 16 to the Golgi apparatus and endoplasmic reticulum requires γ-secretase activity, MBio 5 (5) (2014). p. e01777-14.
[31] I. Aydin, et al., Large scale RNAi reveals the requirement of nuclear envelope breakdown for nuclear import of human papillomaviruses, PLoS Pathog. 10 (5) (2014), e1004162.
[32] P.M. Day, et al., The papillomavirus minor capsid protein, L2, induces localization of the major capsid protein, L1, and the viral transcription/replication protein, E2, to PML oncogenic domains, J. Virol. 72 (1) (1998) 142–150.
[33] P.M. Day, et al., Establishment of papillomavirus infection is enhanced by promyelocytic leukemia protein (PML) expression, Proc. Natl. Acad. Sci. U. S. A. 101 (39) (2004) 14252–14257.
[34] R.D. Everett, The spatial organization of DNA virus genomes in the nucleus, PLoS Pathog. 9 (6) (2013), e1003386.
[35] A.A. McBride, Chapter 4 replication and partitioning of papillomavirus genomes, in: Advances in Virus Research, Academic Press, 2008, pp. 155–205.
[36] M. Leto, et al., Human papillomavirus infection: etiopathogenesis, molecular biology and clinical manifestations, An. Bras. Dermatol. 86 (2) (2011) 306–317.
[37] V.G. Wilson, et al., Papillomavirus E1 proteins: form, function, and features, Virus Genes 24 (3) (2002) 275–290.
[38] C.M. Sanders, A. Stenlund, Recruitment and loading of the E1 initiator protein: an ATP-dependent process catalysed by a transcription factor, EMBO J. 17 (23) (1998) 7044–7055.
[39] T.R. Sarafi, A.A. McBride, Domains of the BPV-1 E1 replication protein required for origin-specific DNA binding and interaction with the E2 transactivator, Virology 211 (2) (1995) 385–396.
[40] G.A. Maglennon, P. McIntosh, J. Doorbar, Persistence of viral DNA in the epithelial basal layer suggests a model for papillomavirus latency following immune regression, Virology 414 (2) (2011) 153–163.
[41] M. Dreer, S. van de Poel, F. Stubenrauch, Control of viral replication and transcription by the papillomavirus E8^E2 protein, Virus Res. 231 (2017) 96–102.
[42] M.M. Donaldson, W. Boner, I.M. Morgan, TopBP1 regulates human papillomavirus type 16 E2 interaction with chromatin, J. Virol. 81 (8) (2007) 4338–4342.
[43] J.A. Westrich, C.J. Warren, D. Pyeon, Evasion of host immune defenses by human papillomavirus, Virus Res. 231 (2017) 21–33.
[44] J.A. Smith, et al., SMCX and components of the TIP60 complex contribute to E2 regulation of the HPV E6/E7 promoter, Virology 468-470 (2014) 311–321.
[45] J.A. Smith, et al., Genome-wide siRNA screen identifies SMCX, EP400, and Brd4 as E2-dependent regulators of human papillomavirus oncogene expression, Proc. Natl. Acad. Sci. 107 (8) (2010) 3752–3757.
[46] M. Oldak, et al., The human papillomavirus type 8 E2 protein suppresses beta4-integrin expression in primary human keratinocytes, J. Virol. 78 (19) (2004) 10738–10746.
[47] N. Egawa, et al., Human papillomaviruses; epithelial tropisms, and the development of neoplasia, Viruses 7 (7) (2015) 3863–3890.
[48] S.B. Vande Pol, A.J. Klingelhutz, Papillomavirus E6 oncoproteins, Virology 445 (1–2) (2013) 115–137.

[49] A. Roman, K. Munger, The papillomavirus E7 proteins, Virology 445 (1–2) (2013) 138–168.
[50] J.T. Thomas, et al., Human papillomavirus type 31 oncoproteins E6 and E7 are required for the maintenance of episomes during the viral life cycle in normal human keratinocytes, Proc. Natl. Acad. Sci. U. S. A. 96 (15) (1999) 8449–8454.
[51] H.K. Wang, et al., Robust production and passaging of infectious HPV in squamous epithelium of primary human keratinocytes, Genes Dev. 23 (2) (2009) 181–194.
[52] D. Patel, et al., The E6 protein of human papillomavirus type 16 binds to and inhibits co-activation by CBP and p300, EMBO J. 18 (18) (1999) 5061–5072.
[53] H. Zimmermann, et al., The human papillomavirus type 16 E6 oncoprotein can down-regulate p53 activity by targeting the transcriptional coactivator CBP/p300, J. Virol. 73 (8) (1999) 6209–6219.
[54] F. Mantovani, L. Banks, The human papillomavirus E6 protein and its contribution to malignant progression, Oncogene 20 (54) (2001) 7874–7887.
[55] S.G. Hwang, et al., Human papillomavirus type 16 E7 binds to E2F1 and activates E2F1-driven transcription in a retinoblastoma protein-independent manner, J. Biol. Chem. 277 (4) (2002) 2923–2930.
[56] W.K. Songock, S.-M. Kim, J.M. Bodily, The human papillomavirus E7 oncoprotein as a regulator of transcription, Virus Res. 231 (2017) 56–75.
[57] K. Van Doorslaer, Z. Li, S. Xirasagar, P. Maes, D. Kaminsky, D. Liou, Q. Sun, R. Kaur, Y. Huyen, A.A. McBride, The papillomavirus episteme: a major update to the papillomavirus sequence database, Nucleic Acids Res. 45 (D1) (2017) D499–D506.
[58] E.L. Prescott, et al., Human papillomavirus type 1 E1^E4 protein is a potent inhibitor of the serine-arginine (SR) protein kinase SRPK1 and inhibits phosphorylation of host SR proteins and of the viral transcription and replication regulator E2, J. Virol. 88 (21) (2014) 12599–12611.
[59] J. Doorbar, et al., The E1E4 protein of human papillomavirus type 16 associates with a putative RNA helicase through sequences in its C terminus, J. Virol. 74 (21) (2000) 10081–10095.
[60] Q. Wang, et al., Functional analysis of the human papillomavirus type 16 E1=E4 protein provides a mechanism for in vivo and in vitro keratin filament reorganization, J. Virol. 78 (2) (2004) 821–833.
[61] J.T. Bryan, D.R. Brown, Association of the human papillomavirus type 11 E1∧E4 protein with Cornified cell envelopes derived from infected genital epithelium, Virology 277 (2) (2000) 262–269.
[62] J. Doorbar, et al., Specific interaction between HPV-16 E1-E4 and cytokeratins results in collapse of the epithelial cell intermediate filament network, Nature 352 (6338) (1991) 824–827.
[63] G.H. Ashrafi, et al., E5 protein of human papillomavirus 16 downregulates HLA class I and interacts with the heavy chain via its first hydrophobic domain, Int. J. Cancer 119 (9) (2006) 2105–2112.
[64] D. DiMaio, L.M. Petti, The E5 proteins, Virology 445 (1–2) (2013) 99–114.
[65] A.N. Blackford, S.P. Jackson, ATM, ATR, and DNA-PK: the trinity at the heart of the DNA damage response, Mol. Cell 66 (6) (2017) 801–817.
[66] C.A. Moody, L.A. Laimins, Human papillomaviruses activate the ATM DNA damage pathway for viral genome amplification upon differentiation, PLoS Pathog. 5 (10) (2009) e1000605.
[67] N.S. Banerjee, et al., Human papillomavirus (HPV) E7 induces prolonged G2 following S phase reentry in differentiated human keratinocytes, J. Biol. Chem. 286 (17) (2011) 15473–15482.
[68] A. Fradet-Turcotte, et al., Nuclear accumulation of the papillomavirus E1 helicase blocks S-phase progression and triggers an ATM-dependent DNA damage response, J. Virol. 85 (17) (2011) 8996–9012.

[69] T. Reinson, et al., Engagement of the ATR-dependent DNA damage response at the human papillomavirus 18 replication centers during the initial amplification, J. Virol. 87 (2) (2013) 951–964.
[70] N. Sakakibara, R. Mitra, A.A. McBride, The papillomavirus E1 helicase activates a cellular DNA damage response in viral replication foci, J. Virol. 85 (17) (2011) 8981–8995.
[71] A.A. McBride, The papillomavirus E2 proteins, Virology 445 (1–2) (2013) 57–79.
[72] T. Iftner, et al., Involvement of Brd4 in different steps of the papillomavirus life cycle, Virus Res. 231 (2017) 76–82.
[73] T. Yu, Y.-C. Peng, E.J. Androphy, Mitotic kinesin-like protein 2 binds and colocalizes with papillomavirus E2 during mitosis, J. Virol. 81 (4) (2007) 1736–1745.
[74] J.L. Parish, et al., ChlR1 is required for loading papillomavirus E2 onto mitotic chromosomes and viral genome maintenance, Mol. Cell 24 (6) (2006) 867–876.
[75] M.A. Stanley, Epithelial cell responses to infection with human papillomavirus, Clin. Microbiol. Rev. 25 (2) (2012) 215–222.
[76] S.V. Graham, Human papillomavirus: gene expression, regulation and prospects for novel diagnostic methods and antiviral therapies, Future Microbiol. 5 (10) (2010) 1493–1506.
[77] C. Johansson, S. Schwartz, Regulation of human papillomavirus gene expression by splicing and polyadenylation, Nat. Rev. Microbiol. 11 (4) (2013) 239–251.
[78] M.G. McPhillips, et al., SF2/ASF binds the human papillomavirus type 16 late RNA control element and is regulated during differentiation of virus-infected epithelial cells, J. Virol. 78 (19) (2004) 10598–10605.
[79] S.A. Cumming, et al., The RNA stability regulator HuR regulates L1 protein expression in vivo in differentiating cervical epithelial cells, Virology 383 (1) (2009) 142–149.
[80] S.A. Cumming, et al., Activity of the human papillomavirus type 16 late negative regulatory element is partly due to four weak consensus 5′ splice sites that bind a U1 snRNP-like complex, J. Virol. 77 (9) (2003) 5167–5177.
[81] T. Cheunim, et al., The alternative splicing factor hnRNP A1 is up-regulated during virus-infected epithelial cell differentiation and binds the human papillomavirus type 16 late regulatory element, Virus Res. 131 (2) (2008) 189–198.
[82] K.A. Becker, et al., Dissection of human papillomavirus type 33 L2 domains involved in nuclear domains (ND) 10 homing and reorganization, Virology 314 (1) (2003) 161–167.
[83] L. Florin, et al., Assembly and translocation of papillomavirus capsid proteins, J. Virol. 76 (19) (2002) 10009–10014.
[84] Y. Stauffer, et al., Infectious human papillomavirus type 18 pseudovirions, J. Mol. Biol. 283 (3) (1998) 529–536.
[85] P.N. Deo, R. Deshmukh, Pathophysiology of keratinization, J. Oral Maxillofac. Pathol. 22 (1) (2018) 86–91.
[86] A.W. Stadnyk, Cytokine production by epithelial cells, FASEB J. 8 (13) (1994) 1041–1047.
[87] B. Wang, P. Amerio, D.N. Sauder, Role of cytokines in epidermal Langerhans cell migration, J. Leukoc. Biol. 66 (1) (1999) 33–39.
[88] M. Scott, M. Nakagawa, A.B. Moscicki, Cell-mediated immune response to human papillomavirus infection, Clin. Diagn. Lab. Immunol. 8 (2) (2001) 209–220.
[89] L. Braun, et al., Regulation of growth and gene expression in human papillomavirus-transformed keratinocytes by transforming growth factor-beta: implications for the control of papillomavirus infection, Mol. Carcinog. 6 (2) (1992) 100–111.
[90] L. Braun, et al., Differential response of nontumorigenic and tumorigenic human papillomavirus type 16-positive epithelial cells to transforming growth factor beta 1, Cancer Res. 50 (22) (1990) 7324–7332.

[91] C.D. Woodworth, V. Notario, J.A. DiPaolo, Transforming growth factors beta 1 and 2 transcriptionally regulate human papillomavirus (HPV) type 16 early gene expression in HPV-immortalized human genital epithelial cells, J. Virol. 64 (10) (1990) 4767–4775.
[92] M.K. Shier, et al., Correlation of TGF beta 1 overexpression with down-regulation of proliferation-inducing molecules in HPV-11 transformed human tissue xenografts, Anticancer Res 19 (6b) (1999) 4969–4976.
[93] M.K. Shier, et al., Transforming growth factor beta 1 (TGF beta 1) down-regulates expression and function of proliferation-inducing molecules in HPV-transformed cells, Anticancer Res 19 (6b) (1999) 4977–4982.
[94] C.D. Woodworth, et al., Transforming growth factor beta 1 supports autonomous growth of human papillomavirus-immortalized cervical keratinocytes under conditions promoting squamous differentiation, Cell Growth Differ. 7 (6) (1996) 811–820.
[95] J. Malejczyk, et al., Autocrine growth limitation of human papillomavirus type 16-harboring keratinocytes by constitutively released tumor necrosis factor-alpha, J. Immunol. 149 (8) (1992) 2702–2708.
[96] L.L. Villa, et al., Differential effect of tumor necrosis factor on proliferation of primary human keratinocytes and cell lines containing human papillomavirus types 16 and 18, Mol. Carcinog. 6 (1) (1992) 5–9.
[97] K.B. Vieira, D.J. Goldstein, L.L. Villa, Tumor necrosis factor alpha interferes with the cell cycle of normal and papillomavirus-immortalized human keratinocytes, Cancer Res. 56 (10) (1996) 2452–2457.
[98] S. Kyo, et al., Regulation of early gene expression of human papillomavirus type 16 by inflammatory cytokines, Virology 200 (1) (1994) 130–139.
[99] C.D. Woodworth, et al., Interleukin 1 alpha and tumor necrosis factor alpha stimulate autocrine amphiregulin expression and proliferation of human papillomavirus-immortalized and carcinoma-derived cervical epithelial cells, Proc. Natl. Acad. Sci. U. S. A. 92 (7) (1995) 2840–2844.
[100] M.A. Khan, et al., Inhibition of growth, transformation, and expression of human papillomavirus type 16 E7 in human keratinocytes by alpha interferons, J. Virol. 67 (6) (1993) 3396–3403.
[101] J. Bekisz, et al., Antiproliferative properties of type I and type II interferon, Pharmaceuticals (Basel, Switzerland) 3 (4) (2010) 994–1015.
[102] A. Nawa, et al., Selective suppression of human papilloma virus type 18 mRNA level in HeLa cells by interferon, Biochem. Biophys. Res. Commun. 170 (2) (1990) 793–799.
[103] S.E. Perea, et al., Interferon-alpha elicits downregulation of human papillomavirus 18 mRNA in HeLa cells by selective repression of endogenous viral transcription, J. Interferon Cytokine Res. 15 (6) (1995) 495–501.
[104] C.D. Woodworth, et al., Leukoregulin and gamma-interferon inhibit human papillomavirus type 16 gene transcription in human papillomavirus-immortalized human cervical cells, Cancer Res. 52 (2) (1992) 456–463.
[105] E. De Marco, M.L. Marcante, Cellular and molecular analyses of interferon beta cytopathic effect on HPV-16 in vitro transformed human keratinocytes (HPK-IA), J. Biol. Regul. Homeost. Agents 9 (1) (1995) 24–30.
[106] F. De Marco, F. Giannoni, M.L. Marcante, Interferon-beta strong cytopathic effect on human papillomavirus type 16-immortalized HPK-IA cell line, unexpectedly not shared by interferon-alpha, J. Gen. Virol. 76 (Pt 2) (1995) 445–450.
[107] K. Iwatsuki, et al., Plane warts under spontaneous regression. Immunopathologic study on cellular constituents leading to the inflammatory reaction, Arch. Dermatol. 122 (6) (1986) 655–659.
[108] O.M. Memar, I. Arany, S.K. Tyring, Skin-associated lymphoid tissue in human immunodeficiency virus-1, human papillomavirus, and herpes simplex virus infections, J. Invest. Dermatol. 105 (1 Suppl) (1995) 99s–104s.

[109] A.E. Morelli, et al., Cellular subsets and epithelial ICAM-1 and HLA-DR expression in human papillomavirus infection of the vulva, Acta Derm.Venereol. 74 (1) (1994) 45–50.
[110] J. Viac, et al., Expression of immune associated surface antigens of keratinocytes in human papillomavirus-derived lesions, Immunobiology 188 (4–5) (1993) 392–402.
[111] H.H. Morris, et al., Langerhans' cells in human cervical epithelium: effects of wart virus infection and intraepithelial neoplasia, Br. J. Obstet. Gynaecol. 90 (5) (1983) 412–420.
[112] S.K. Tay, et al., Subpopulations of Langerhans' cells in cervical neoplasia, Br. J. Obstet. Gynaecol. 94 (1) (1987) 10–15.
[113] M.B. Lappin, I. Kimber, M. Norval, The role of dendritic cells in cutaneous immunity, Arch. Dermatol. Res. 288 (3) (1996) 109–121.
[114] H.A. Cubie, et al., Lymphoproliferative response to fusion proteins of human papillomaviruses in patients with cervical intraepithelial neoplasia, Epidemiol. Infect. 103 (3) (1989) 625–632.
[115] T.D. de Gruijl, et al., Differential T helper cell responses to human papillomavirus type 16 E7 related to viral clearance or persistence in patients with cervical neoplasia: a longitudinal study, Cancer Res. 58 (8) (1998) 1700–1706.
[116] M. Nakagawa, et al., T-cell proliferative response to human papillomavirus type 16 peptides: relationship to cervical intraepithelial neoplasia, Clin. Diagn. Lab. Immunol. 3 (2) (1996) 205–210.
[117] F.X. Bosch, et al., Prevalence of human papillomavirus in cervical cancer: a worldwide perspective. International biological study on cervical cancer (IBSCC) study group, J. Natl. Cancer Inst. 87 (11) (1995) 796–802.
[118] J.C. Luxton, et al., Proliferative T cell responses to the human papillomavirus type 16 E7 protein in women with cervical dysplasia and cervical carcinoma and in healthy individuals, J. Gen.Virol. 77 (Pt 7) (1996) 1585–1593.
[119] A.S. Kadish, et al., Cell-mediated immune responses to E7 peptides of human papillomavirus (HPV) type 16 are dependent on the HPV type infecting the cervix whereas serological reactivity is not type-specific, J. Gen.Virol. 75 (Pt 9) (1994) 2277–2284.
[120] A.S. Kadish, et al., Lymphoproliferative responses to human papillomavirus (HPV) type 16 proteins E6 and E7: outcome of HPV infection and associated neoplasia, J. Natl. Cancer Inst. 89 (17) (1997) 1285–1293.
[121] T.D. de Gruijl, et al., T cell proliferative responses against human papillomavirus type 16 E7 oncoprotein are most prominent in cervical intraepithelial neoplasia patients with a persistent viral infection, J. Gen.Virol. 77 (Pt 9) (1996) 2183–2191.
[122] M.H. Stoler, et al., Human papillomavirus type 16 and 18 gene expression in cervical neoplasias, Hum. Pathol. 23 (2) (1992) 117–128.
[123] J. Cason, et al., Identification of immunogenic regions of the major coat protein of human papillomavirus type 16 that contain type-restricted epitopes, J. Gen.Virol. 70 (11) (1989) 2973–2987.
[124] U. Sahu, et al., Interleukin-17—a multifaceted cytokine in viral infections, J. Cell. Physiol. (2021) 1–2.
[125] U. Sahu, P. Khare, Role of interleukin-17 in human papillomavirus infection and associated malignancies, Microb. Pathog. (2021), 105294.
[126] C. Zhou, Z.K. Tuong, I.H. Frazer, Papillomavirus immune evasion strategies target the infected cell and the local immune system, Front. Oncol. 9 (682) (2019).
[127] L.S. Miller, Toll-like receptors in skin, Adv. Dermatol. 24 (2008) 71–87.
[128] L. Pacini, et al., Downregulation of toll-like receptor 9 expression by Beta human papillomavirus 38 and implications for cell cycle control, J. Virol. 89 (22) (2015) 11396–11405.
[129] U.A. Hasan, et al., The human papillomavirus type 16 E7 oncoprotein induces a transcriptional repressor complex on the toll-like receptor 9 promoter, J. Exp. Med. 210 (7) (2013) 1369–1387.

[130] L. Lau, et al., DNA tumor virus oncogenes antagonize the cGAS-STING DNA-sensing pathway, Science 350 (6260) (2015) 568–571.
[131] I.I. Daud, et al., Association between toll-like receptor expression and human papillomavirus type 16 persistence, Int. J. Cancer 128 (4) (2011) 879–886.
[132] F. Cannella, et al., TLR9 is expressed in human papillomavirus-positive cervical cells and is overexpressed in persistent infections, Immunobiology 220 (3) (2015) 363–368.
[133] J.W. Lee, et al., Increased toll-like receptor 9 expression in cervical neoplasia, Mol. Carcinog. 46 (11) (2007) 941–947.
[134] L.V. Ronco, et al., Human papillomavirus 16 E6 oncoprotein binds to interferon regulatory factor-3 and inhibits its transcriptional activity, Genes Dev. 12 (13) (1998) 2061–2072.
[135] J.S. Park, et al., Inactivation of interferon regulatory factor-1 tumor suppressor protein by HPV E7 oncoprotein. Implication for the E7-mediated immune evasion mechanism in cervical carcinogenesis, J. Biol. Chem. 275 (10) (2000) 6764–6769.
[136] L.B. Ivashkiv, L.T. Donlin, Regulation of type I interferon responses, Nat. Rev. Immunol. 14 (1) (2014) 36–49.
[137] S. Li, et al., The human papilloma virus (HPV)-18 E6 oncoprotein physically associates with Tyk2 and impairs Jak-STAT activation by interferon-alpha, Oncogene 18 (42) (1999) 5727–5737.
[138] Y.E. Chang, L.A. Laimins, Microarray analysis identifies interferon-inducible genes and Stat-1 as major transcriptional targets of human papillomavirus type 31, J. Virol. 74 (9) (2000) 4174–4182.
[139] S. Hong, K.P. Mehta, L.A. Laimins, Suppression of STAT-1 expression by human papillomaviruses is necessary for differentiation-dependent genome amplification and plasmid maintenance, J. Virol. 85 (18) (2011) 9486–9494.
[140] P. Barnard, N.A. McMillan, The human papillomavirus E7 oncoprotein abrogates signaling mediated by interferon-alpha, Virology 259 (2) (1999) 305–313.
[141] P. Barnard, E. Payne, N.A. McMillan, The human papillomavirus E7 protein is able to inhibit the antiviral and anti-growth functions of interferon-alpha, Virology 277 (2) (2000) 411–419.
[142] R. Karim, et al., Human papillomavirus (HPV) upregulates the cellular deubiquitinase UCHL1 to suppress the keratinocyte's innate immune response, PLoS Pathog. 9 (5) (2013), e1003384.
[143] E.R. Vandermark, et al., Human papillomavirus type 16 E6 and E 7 proteins alter NF-kB in cultured cervical epithelial cells and inhibition of NF-kB promotes cell growth and immortalization, Virology 425 (1) (2012) 53–60.
[144] S.M. Huang, D.J. McCance, Down regulation of the interleukin-8 promoter by human papillomavirus type 16 E6 and E7 through effects on CREB binding protein/p300 and P/CAF, J. Virol. 76 (17) (2002) 8710–8721.
[145] C. Heller, et al., Identification of key amino acid residues that determine the ability of high risk HPV16-E7 to dysregulate major histocompatibility complex class I expression, J. Biol. Chem. 286 (13) (2011) 10983–10997.
[146] H. Li, et al., HPV16E7 mediates HADC chromatin repression and downregulation of MHC class I genes in HPV16 tumorigenic cells through interaction with an MHC class I promoter, Biochem. Biophys. Res. Commun. 349 (4) (2006) 1315–1321.
[147] N.T. Georgopoulos, J.L. Proffitt, G.E. Blair, Transcriptional regulation of the major histocompatibility complex (MHC) class I heavy chain, TAP1 and LMP2 genes by the human papillomavirus (HPV) type 6b, 16 and 18 E7 oncoproteins, Oncogene 19 (42) (2000) 4930–4935.
[148] M.S. Cortese, G.H. Ashrafi, M.S. Campo, All 4 di-leucine motifs in the first hydrophobic domain of the E5 oncoprotein of human papillomavirus type 16 are essential for surface MHC class I downregulation activity and E5 endomembrane localization, Int. J. Cancer 126 (7) (2010) 1675–1682.

[149] M. Gruener, et al., The E5 protein of the human papillomavirus type 16 down-regulates HLA-I surface expression in calnexin-expressing but not in calnexin-deficient cells, Virol. J. 4 (1) (2007) 116.

[150] F.V. Cromme, et al., Differences in MHC and TAP-1 expression in cervical cancer lymph node metastases as compared with the primary tumours, Br. J. Cancer 69 (6) (1994) 1176–1181.

[151] H.J. Bontkes, et al., Specific HLA class I down-regulation is an early event in cervical dysplasia associated with clinical progression, Lancet 351 (9097) (1998) 187–188.

[152] K.H. Kim, et al., CD4+ T-cell response against human papillomavirus type 16 E6 protein is associated with a favorable clinical trend, Cancer Immunol. Immunother. 61 (1) (2012) 63–70.

[153] J.W. Molling, et al., CD4+CD25hi regulatory T-cell frequency correlates with persistence of human papillomavirus type 16 and T helper cell responses in patients with cervical intraepithelial neoplasia, Int. J. Cancer 121 (8) (2007) 1749–1755.

[154] P.J. de Vos van Steenwijk, et al., Surgery followed by persistence of high-grade squamous intraepithelial lesions is associated with the induction of a dysfunctional HPV16-specific T-cell response, Clin. Cancer Res. 14 (22) (2008) 7188–7195.

[155] S.H. van der Burg, et al., Association of cervical cancer with the presence of CD4+ regulatory T cells specific for human papillomavirus antigens, Proc. Natl. Acad. Sci. 104 (29) (2007) 12087–12092.

[156] C.B. Buck, P.M. Day, B.L. Trus, The papillomavirus major capsid protein L1, Virology 445 (1–2) (2013) 169–174.

[157] S. DiGiuseppe, et al., The nuclear retention signal of HPV16 L2 protein is essential for incoming viral genome to transverse the trans-Golgi network, Virology 458-459 (2014) 93–105.

[158] J.G. Joyce, et al., The L1 major capsid protein of human papillomavirus type 11 recombinant virus-like particles interacts with heparin and cell-surface Glycosaminoglycans on human keratinocytes*, J. Biol. Chem. 274 (9) (1999) 5810–5822.

[159] F. Stubenrauch, H.B. Lim, L.A. Laimins, Differential requirements for conserved E2 binding sites in the life cycle of oncogenic human papillomavirus type 31, J. Virol. 72 (2) (1998) 1071–1077.

[160] H.L. Howie, R.A. Katzenellenbogen, D.A. Galloway, Papillomavirus E6 proteins, Virology 384 (2) (2009) 324–334.

[161] D. Zhu, M. Ye, W. Zhang, E6/E7 oncoproteins of high risk HPV-16 upregulate MT1-MMP, MMP-2 and MMP-9 and promote the migration of cervical cancer cells, Int. J. Clin. Exp. Pathol. 8 (5) (2015) 4981–4989.

[162] L.F. Wetherill, et al., High-risk human papillomavirus E5 oncoprotein displays channel-forming activity sensitive to small-molecule inhibitors, J. Virol. 86 (9) (2012) 5341–5351.

[163] J.C. Luxton, et al., Serological and T-helper cell responses to human papillomavirus type 16 L1 in women with cervical dysplasia or cervical carcinoma and in healthy controls, J. Gen. Virol. 78 (Pt 4) (1997) 917–923.

[164] P.S. Shepherd, et al., Proliferative T cell responses to human papillomavirus type 16 L1 peptides in patients with cervical dysplasia, J. Gen. Virol. 77 (Pt 4) (1996) 593–602.

[165] A.L. Glasebrook, F.W. Fitch, T-cell lines which cooperate in generation of specific cytolytic activity, Nature 278 (5700) (1979) 171–173.

[166] B.J. Alter, et al., Secondary cell-mediated lympholysis: importance of H-2 LD and SD factors, J. Exp. Med. 143 (5) (1976) 1005–1014.

[167] A.L. DeFranco, Molecular aspects of B-lymphocyte activation, Annu. Rev. Cell Biol. 3 (1987) 143–178.

[168] R.J. Noelle, E.C. Snow, T helper cell-dependent B cell activation, FASEB J. 5 (13) (1991) 2770–2776.

CHAPTER 3

Malignancies associated with HPV

Prashant Kumar Tiwari[a], Divya Kushwaha[b], and Ambuj Kumar Kushwaha[a]
[a]School of Biotechnology, Institute of Science, Banaras Hindu University, Varanasi, UP, India
[b]Department of Chemistry, MMV, Banaras Hindu University, Varanasi, UP, India

1. Introduction

Papillomaviruses are a ubiquitous group of DNA viruses causing different clinical manifestations in vertebrates [1]. The malignancies associated with human papillomavirus (HPV) make up about 5% of all cancers that affect more than 600,000 people worldwide each year [1]. Cervical cancer (CC) [99.7%, the fourth most frequent cancer in women] [2], and squamous cell carcinomas (SCC) of the oropharynx, anus, rectum, penis, vagina, and vulva are among the HPV-related cancers (Fig. 1).

HPVs, first described by Shope and Hurst in 1933 [3], are a large family of viruses carrying small circular double-stranded DNA within a capsid that displays icosahedral symmetry consisting of 72 star-shaped pentamer capsomeres of approximately 60 nm in size. Infected cells contain the viral genome in the form of a superhelix with histoproteins generated from the host cell [4]. HPVs are primarily transmitted via direct physical contact with an HPV-infected individual, for instance, skin to skin, skin to the mucosa, and mucosa to mucosa [5]. The causes of HPV infection may include sexual activity, multiple sex partners, use of oral contraceptives, and history of prior infection [6]. HPV infections can be asymptomatic and eliminated by the immune system, but they can also progress to persistent HPV infections, which can lead to intraepithelial neoplasia, a low- or high-grade precancerous condition, and eventually malignancy. The dysregulated inflammatory response induced by HPV infection helps in viral persistence and malignant progression [7,8].

More than 200 HPV genotypes have been identified and characterized into five evolutionary genera (α, β, γ, μ, and ν) and numbered species (for example, α-1) (Fig. 2). Remarkably, each genotype of HPV is evolutionarily adapted to infect particular human epithelial tissue such as anogenital or mucosa. Furthermore, each HPV strain has developed its own technique for evading the

Immunopathology, Diagnosis, and Treatment of HPV-induced Malignancies
https://doi.org/10.1016/B978-0-323-91797-1.00005-3

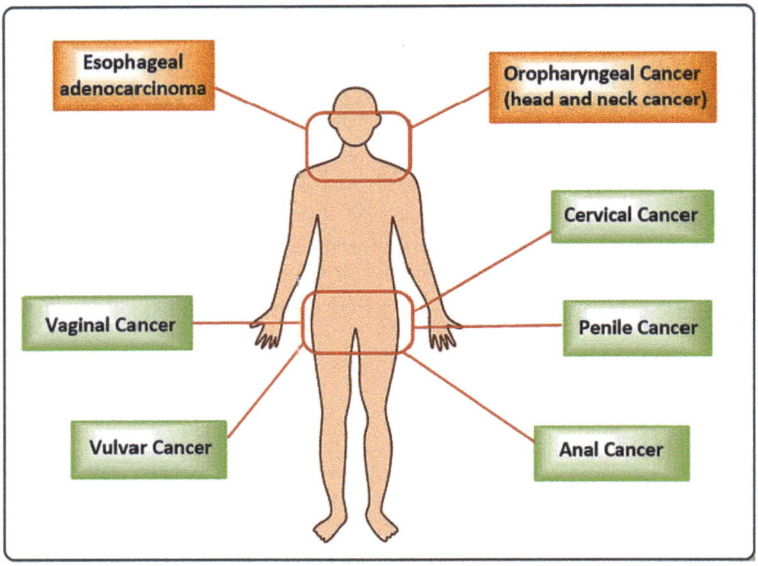

Fig. 1 HPV-induced malignancies in humans. HPV commonly induces oropharyngeal cancer (head and neck cancer), esophageal adenocarcinoma, cervical cancer, vaginal cancer, vulvar cancer, anal cancer, and penile cancer.

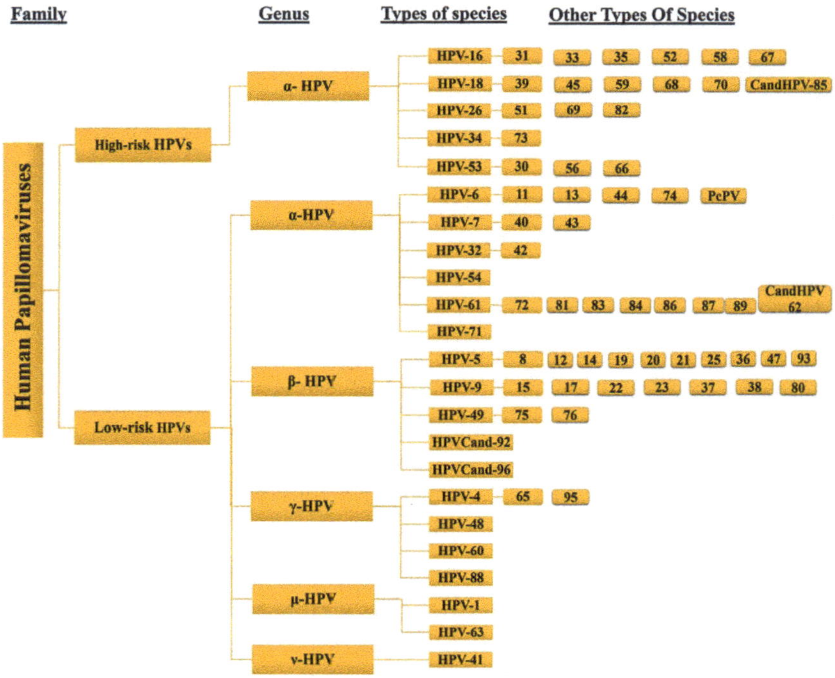

Fig. 2 Taxonomic classification of HPVs developed by the international committee on the taxonomy of viruses.

immune system of the host. For example, to avoid the host's immune reaction, HPV of the genus β reduces viral gene expression. Similarly, HPV of the genus α has developed an immunological escape strategy based on downregulation of viral protein-mediated immune responses. The expression of viral genes differs significantly between these strategies of immune evasion, and more so, the actions of the same viral proteins depend on the type of infected tissues [9].

Based on their clinical significance, HPVs are divided into two groups: low risk of HPV (LR-HPV) genotypes associated with low risk of carcinogenesis, which mainly causes condylomas, and high risk of HPV (HR-HPV) genotypes, which are strongly associated with the development of neoplasia and cancer. The LR-HPV results in genital warts but does not cause malignancies. The HR-HPV is associated with different types of malignancies and include HPV-16, 18, 31, 33, 35, 39, 45, 51, 52, 56, 58, and 59 of which HPV-16 is extremely carcinogenic and most detected HPV genotype at the population level [10–12] (Fig. 3).

As per the World Health Organization (WHO), HPV-16 and HPV-18 are the leading causes of CC in women that together account for ~70% of cervical and precancerous cervical lesions. Moreover, many other diseases like cutaneous warts, anogenital warts, genital and upper aerodigestive tract cancer are associated with HPV infection [13].

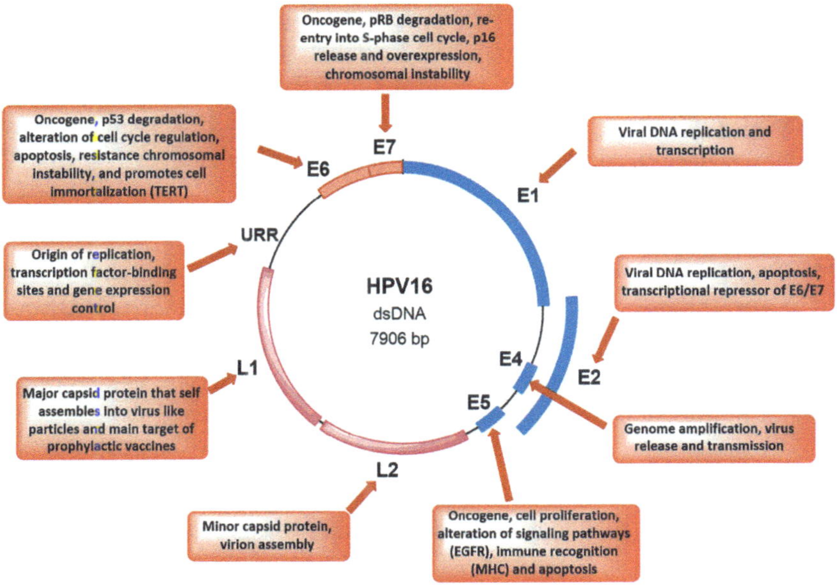

Fig. 3 Structure of HPV 16 genome and viral proteins with their key functions.

2. Malignancies associated with HPV

HPV is associated with both premalignant and malignant lesions, including cervical [14–16], anogenital (penile, anal, vaginal, vulvar), oropharyngeal (head and neck) cancer, and esophageal carcinoma [7,17–20].

2.1 Cervical cancer

Cervical cancer (CC) is the fourth most frequent malignancy in women worldwide, affecting primarily women from 25 to 30 years of age [21]. Infection with HR-HPVs is substantially linked to the development of CC and its precursor lesions, according to epidemiological and molecular studies [22,23]. HPV oncoproteins E6 and E7 (Fig. 3) are required for the development of CC [24,25].

Intraepithelial neoplasia characterizes HPV-associated precancer states, which are named after the infection site, such as cervical intraepithelial neoplasia (CIN), vulval intraepithelial neoplasia (VIN), vaginal intraepithelial neoplasia (VAIN), penile intraepithelial neoplasia (PIN), and abdominal anal intraepithelial neoplasia (AIN). CIN can range in severity from CIN1 to CIN3 [26–28]. HPV-16 and HPV-33 were found to be linked with the highest risk of CIN formation in the HPV vaccination immunogenicity and efficacy (VIVIANE) research, followed by HPV-18, HPV-31, and HPV-45 [27]. HPV-16 viral loads can be used to determine the severity of HPV-associated cervical neoplasia, as multiple studies have found a link between an increase in HPV-16 viral load and an increase in lesion grade [29,30]. CIN1 is a low-grade squamous intraepithelial lesion (LSIL) that reflects an infection's first stage, whereas CIN2 and CIN3 are high-grade dysplasia or high-grade squamous intraepithelial lesion, respectively [31,32]. CIN2 is less prevalent as disease progresses; however, CIN3 is a real precancerous condition that can lead to invasive cancer [31,33]. CC is divided into two categories based on the type of cells that produce cancer: squamous cell carcinoma and adenocarcinoma. The most common type of CC is squamous cell carcinoma, which arises from cells in the exocervix (outer part of the cervix) [34]. Adenocarcinoma is a less common type of cancer that arises from the endocervical canal's glandular epithelium.

2.1.1 Epidemiology and pathology

CC is one of the most common cancers in women, with approximately 570,000 new cases in 2018 and 7.5% of all female cancer fatalities (more than 250,000) worldwide [35,36]. Most of the CCs are HPV-associated, such as cervical squamous cell carcinomas (70%), cervical adenocarcinomas (25%), and other CCs [37,38]. CC mortality varies depending on the country's economic levels. CC is substantially less common in affluent countries, with an incidence rate of 5.0 per 100,000, compared to 8.0 per 100,000 in underdeveloped countries [39]. CC mortality is also lower in developed countries, at 2.2 per 100,000, compared to 4.3 per 100,000 in developing countries. Furthermore, the prevalence of CC varies by geographic region. In 2012, there were 34.8 new cases and 22.5 fatalities per 100,000 women in sub-Saharan Africa, but only 4.4 new cases and 1.9 deaths per 100,000 women in Western Asia. In contrast, Northern America is reported to have the world's third-lowest CC rate [40]. CC progresses through a complex method of unrestrained cell division that includes HPV gene integration as well as other cellular alterations and epigenetic variables such DNA methylation, histone acetylation, and chromatin modifications. The HPV-16 genome generates benign and precancerous cervix lesions when it is unintegrated (episome), but when it is integrated into the host genome, it can induce CIN3 and CC. Because viral genome integration affects E2 protein, resulting in dysregulation of E6 and E7 gene expression, E2 protein is critical in the carcinogenic process as a regulator of oncoproteins E6 and E7 (oncogenes) [41–43]. Furthermore, epigenetic factors that do not directly alter E2 DNA binding can affect the production of viral E6 & E7 proteins [44].

HPV-associated cancers rely heavily on the expression of the E6 and E7 oncogenes for sustained proliferation and survival [25]. E6 interacts with an E6-associated binding protein (E6AP) ubiquitin-protein ligase, causing structural changes in E6 that allow it to bind and destroy p53 (tumor suppressor protein), resulting in increased cell proliferation. Furthermore, E7 can attach to the retinoblastoma protein pRB, causing it to be inactivated and degraded [45]. Both E6 and E7 have the ability to deregulate microRNA (miRNA), associated with the posttranscriptional control of host genes connected to carcinogenesis [46]. HPV oncoproteins E5, E6, and E7 are reported to control the host miRNA profile in recent research. For example, numerous miRNAs (miR-9, miR-21, miR-155, and miR-203) are overexpressed in HPV-associated CC cells, where miR-9 targets cadherin-1 (CDH-1) and its overexpression is associated with increased cell motility and invasiveness [47], miRNA-21 promotes invasion and cell proliferation by targeting programmed cell death 4 (PDCD4) [48,49], miRNA-155 promotes proliferation by targeting liver kinase B1 (LKB1) [50,51], and miRNA-203 inhibits HPV amplification [52].

Aside from oncoproteins and tumor suppressor genes, epigenetic modulation of host and viral gene expression, which involves changes in DNA methylation, histone modification, and noncoding RNA profile, plays a major role in cancer development. Persistent HPV infection has been linked to CC and has been shown to influence the methylation process of both the host and the viral genome [53].

CC screening programs are helpful in lowering the incidence and death of CC patients. As a result, different CC screening systems have been implemented in various nations. Few nations have launched population-based screening programs in which women from the target group are individually identified and encouraged to participate. Some countries, however, have implemented an opportunistic screening program, in which invites are based on the individual's choice or interactions with healthcare practitioners. Organized CC screening programs, as opposed to opportunistic screening, are demonstrably more effective because they can attain high participation at regular intervals with equal access, as well as high-quality diagnostic criteria [54,55]. Only a few countries, such as Canada, Finland, Australia, the United Kingdom, and Singapore, have organized CC screening programs, whereas opportunistic screening is used in Eastern European countries with low screening and immunization rates, resulting in high incidence of CC and increased mortality rates [56].

The Pap smear test is a widely used tool for detecting minor morphological changes in the cervical epithelium associated with early HPV infections, and it has a sensitivity of 70.80% for detecting a high-grade squamous intraepithelial lesion (HSIL) [57]. Co-screening with an HPV DNA test can improve the sensitivity of early detection of precancerous lesions [58]. Early-stage CC can be treated surgically by radical hysterectomy with pelvic lymphadenectomy. However, in cases where surgical option does not exist, primary radiation therapy (RT) is preferred due to the radiosensitive nature of CCs. Nonetheless, the availability of treatment cannot discount the importance of spreading the CC screening awareness program, educating women about the symptoms of CC, and motivating them to visit the hospital for screening.

2.2 Oropharyngeal cancer

Oropharyngeal cancers (OPCs; head and neck squamous cell carcinomas) are the sixth common cancer type worldwide, accounting for ~5% of all cases. They are the most studied and the most characterized type of cancer. The cases of HPV's association with oral cell squamous cell carcinomas were reported in 2008 for lingual cancer, tonsil cancer, and OPC [59,60]. HPV-associated OPC is primarily linked with oral sex and less frequently with the mutation in the p53 gene [61]. The most prevalent OPC is head and neck squamous cell carcinoma (HNSCC), which arises from the mucosal epithelium of the oral cavity, pharynx, and larynx. Around 25% of cases of HNSCC are caused by HPV infection, with the remaining 75% caused by cigarette and alcohol usage [62]. The association of HPV with OPC was demonstrated by HPV-DNA in oropharyngeal tumors in addition to an increase of HNSCC in patients seropositive for HPV antibodies. Gillison et al. observed that oropharyngeal tumor samples were 7.7 times as likely to have HPV-DNA compared to non-oropharyngeal sites [63]. In most cases, HPV-16 is the major cause of HPV-associated HNSCC followed by HPV-18 and other strains [64]. Mork et al. reported >14 times high risk of developing HNSCC in HPV-16-positive patients than HPV-16-negative patients [65]. Along similar lines, Smith et al. observed that detection of HPV-16 E6 and E7 antibodies signifies 73 times more risk of HNSCC than HPV-16 E6 and E7 negative patients [66].

2.2.1 Epidemiology and pathology

HNSCC was the sixth most frequent cancer type in the globe in 2018, with 450,000 fatalities and 890,000 new cases. It is expected to rise by 30% by 2030 (i.e., 1.08 million new cases each year) [2,67]. Due to the use of particular carcinogen-containing items, risk factors for HNSCC are greater in geographical areas such as Australia and Southeast Asia, but the likelihood of HPV-associated HNSCC is higher in regions such as the United States and Western Europe [68–70]. According to a recent study based on HNSCC patients primarily from Europe and South America, HPV infections were linked to 22.4% OPC, 4% oral cavity cancers, and 3.5% laryngeal cancers among all HNSCC cases examined [71]. Approximately 71% of OPCs in the United States have been linked to HPV infection [72,73].

HNSCC can damage the mucosal epithelial lining of the oral cavity and sinonasal tract. Starting with epithelial cell hyperplasia, invasive HNSCC progresses through dysplasia (mild, moderate, and severe), carcinoma in situ, and finally invasive malignancy [74]. CDKN2A and TP53 (encodes p53) tumor suppressor genes are inactivated in the early stages of HNSCC, while PTEN (phosphatase and tensin homologue) is involved in the later stages. CDKN2A encodes p16INK4A that prevents phosphorylation of RB1 by inhibiting CDK4 and CDK6. HNSCC tumors mostly exhibit frequent mutation of TP53 (72% of tumors) and CDKN2A (22% of tumors) [75].

Currently, there is no effective screening program for HPV-associated HNSCC. A Pap-test equivalent method involving the collection of oropharyngeal cell samples by a cytology brush can be useful for the detection of invasive HNSCC [76]. HNSCC caused by HPV does not have a gross precursor lesion. As a result, co-screening with oral HPV DNA or anti-HPV-16 E6 seropositivity is a more effective and sensitive means of detecting oropharyngeal cancer. Anti-L1 capsid antibodies, which can inhibit the first infection phase of viral entry, are produced by the existing vaccinations. Vaccines are thus only effective for primary prevention.

2.3 Vulvar cancer/vulvar squamous cell carcinoma

Vulvar squamous cell carcinoma (VSCC) is the fourth most common gynecologic carcinoma [77,78]. It can manifest itself in two ways: HPV-dependent and HPV-independent. HSIL or VIL are terms used to describe VSCC caused by HPV (VIN2 & VIN3) [79–81]. VSCC is most commonly associated with HPV-16; however, other HPVs such as HPV-18, HPV-31, HPV-33, and HPV-45 are also frequent [82].

2.3.1 Epidemiology and pathology

Each year vulvar cancer affects almost 27,000 women worldwide, and 88% of the cases are diagnosed with VSCC [83]. VSCC can be distinguished into two different types. The first type involves HPV infection that causes VIN leading to vulvar cancer and the second type of VSCC includes vulvar non-neoplastic epithelial disorders (VNEDs) that lead to cellular atypia and eventually to cancer [84]. Histologically, HPV-associated VSCCs in young patients develop basaloid or warty carcinomas that are usually derived by a precursor called usual-type vulvar intraepithelial neoplasia (uVIN), whereas HPV-independent VSCCs are associated with old age and usually occur in postmenopausal women characterized by typical differentiated keratinized carcinomas derived from a premalignant lesion known as differentiated VIN (dVIN) [85].

2.4 Penile cancer

Penile cancer is a rare malignancy with an incidence rate of 0.8 per 100,000 men worldwide and is predominant in elderly men aged between 50 to 70 years [67]. Its prevalence varies with geographical location and economic disparity as evident by its high incidence in less developed countries as compared to developed Western countries [86,87].

Penile cancer mainly originates in the squamous mucosa of the glans, coronal sulcus, or inner surface of the foreskin of the penile. SCC is responsible for 99% of penile cancer, and about 50% of them are related to HPV [88]. Seropositivity to HPV is more frequent in penile carcinoma [59,89,90], and many epidemiological studies suggest HPV-16 infection associated with penile cancer [91].

2.4.1 Epidemiology and pathology

Penile cancer affects men in Africa, Asia, and South America, accounting for about 10% of the malignant disease in certain of these nations. For example, the incidence rates in Paraguay and Uganda are roughly 4.2 and 4.4 per 100,000 men, respectively [92,93]. In Sweden, approximately 150 new cases are diagnosed each year, with a prevalence of approximately 2 per 100,000 men [88]. The highest frequency of penile cancer in the United States is seen among Hispanics and in the South, as well as in communities with a high level of poverty [87].

Penile cancer follows two distinct pathways: HPV-dependent and HPV-independent penile carcinogenesis. The HPV-dependent penile carcinogenesis develops from the progression of precursor lesions caused by HR-HPV infection, which is similar to the disease progression pathways of HR-HPV-associated cervical carcinogenesis [94]. As in other HPV-positive cells, the E6 & E7 oncogenes are expressed that enhance cell proliferation and malignant transformation leading to penile cancer [95,96].

HPV infection can cause two different types of exterior genital lesions, depending upon the extent of the viral infection of the squamous epithelium: genital warts (*Condyloma acuminatum*) and undifferentiated penile intraepithelial neoplasia (PIN). Genital warts are low-grade lesions caused by virus generation in the squamous epithelium because of a temporary HPV viral infection. The PIN, on the other hand, is HPV-associated precancerous lesions that originate from the viral genome being integrated into the host genome, resulting in oncogene overexpression and malignant transformation [96] (Fig. 4).

The integration of the viral genome occurs at the locus (FAM92A1 gene) of chromosome 8q21.3 and the 16p13.3 locus in HPV-16-mediated penile cancer [TNF receptor-associated protein 1 (TRAP1) gene]. This HPV integration represents an endpoint in cell clonal selection, resulting in changes to functional genes that cause uncontrolled proliferation and dedifferentiation, which can lead to invasive cancer [97]. HPV-associated

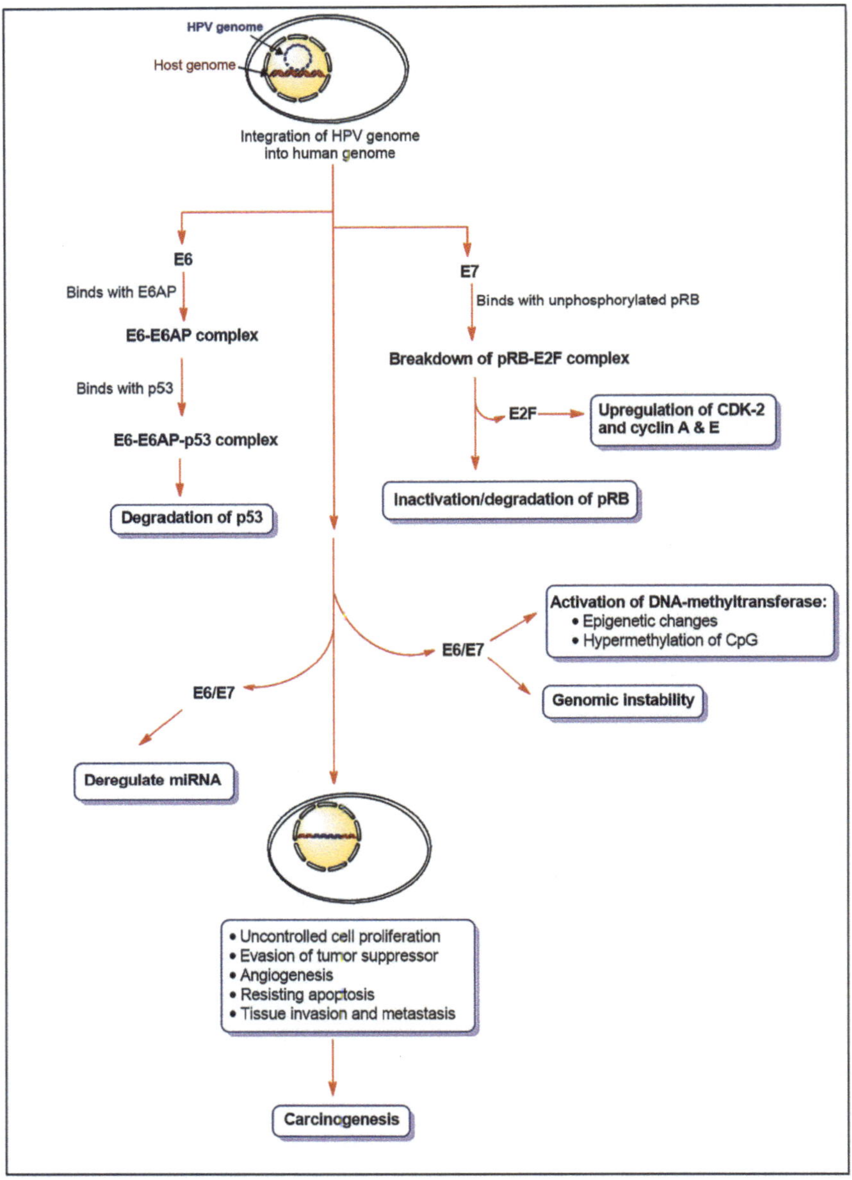

Fig. 4 Progression of carcinogenesis involves HPV genome integration into the host genome resulting in overexpression of oncogenes (E6 & E7). The oncogenes can cause dysregulation in various pathways, for instance, inactivation and deprivation of pRB and p53 that promote uncontrolled cell proliferation, tumor suppressor evasion, angiogenesis, tissue invasion, and metastasis eventually leading to carcinogenesis.

PINs are categorized into low (PIN1), moderate (PIN2), and severe (PIN3), and the chances of occurrence of HPV DNA in PIN1, PIN2, and PIN3 are 75%, 93%, and 100%, respectively [98].

2.5 Esophageal cancer

Esophageal squamous cell carcinoma (ESCC) is the eighth most common cancer worldwide [99,100], where 22.2% of esophageal cancers are associated with HR-HPV, in particular HPV-16 [101,102].

2.5.1 Epidemiology and pathology

The highest frequencies of ESCC may be found in two geographical belts that run from North Central China through the Central Asian republics to Northern Iran, as well as from East to South Africa. More than half of all ESCC-related mortalities occur in the Republic of China [103].

In the upper and middle third of the esophagus, adenocarcinoma of the distal esophageal carcinoma is more prevalent than squamous cell carcinoma. Esophageal squamous cell carcinoma (ESCC) and esophageal adenocarcinoma (EAC) have different geographic patterns, time trends, and important risk factors. In China, the prevalence of EAC is low, ranging from 1.5% to 4.5%. In the United States, however, the rate of EAC among white men has increased from 0.4 per 100,000 men in 1973 to 2.8 per 100,000 men in 2012. A similar pattern was discovered in the Northern Europe, Australia, and United Kingdom. In these countries, the ratio of ESCC to EAC among white men was 4.7:1 in 1975 and 0.43:1 in 1996–98 [104]. During this time, the incidence of EAC increased by 8%–10% per year, albeit there has been a recent decrease in this trend [105]. The large increase in the frequency of EAC in Western countries [106] and a few developing countries corresponds to an increased prevalence of gastroesophageal reflux disease (GERD) and obesity, both of which are recognized risk factors for EAC [107]. In esophageal cancer biopsies, the DNA of HPV-11 is predominant. However, DNA of other HPV genotypes (HPV-6, HPV-11, HPV-16, and HPV-52) has also been detected [108,109].

2.6 Vaginal cancer

Primary vaginal cancer is rare cancer. Every year, approximately 13,200 women are diagnosed with vaginal cancer worldwide [83]. Vaginal cancer accounts for about 1%–2% of all gynecological cancers in the United States and about 0.2 to 0.7 per 100,000 newly diagnosed cases per year in Asia [105,110,111]. The incidence of VAIN is much lower in comparison to the incidence ofVIN [112]. Up to 80% of reported initial vaginal cancers are squamous cell carcinomas, while about 14% of primary vaginal cancers are histology adenocarcinomas, which typically affect younger women [113,114].

2.6.1 Epidemiology and pathology

The most prevalent type of HPV identified in vaginal malignancies is HPV-16 [115,116]. HPV DNA is found in about 60% of invasive and 80%–90% of in situ vaginal squamous cell carcinomas [117,118]. Seropositivity to HPV-18 doubles the risk of in situ cancers, while antibodies to HPV-16 raise the risk of invasive carcinoma by up to sixfold [115,118].About 30% of individuals with primary vaginal cancer have had a CIN or invasive CC for at least 5 years prior to diagnosis [119].Women with malignancies at other anogenital sites [120] or a family history of anogenital cancer [117,121] have been linked to an increased risk of vaginal cancer.

2.7 Anal cancer

Anal cancer is another rare cancer in the general population worldwide with age-adjusted annual incidence rates of less than 2 per 100,000 cases [122]. In general, anal cancer is more prevalent in women than in men; however, gender-based variation in the number of cases has been observed with few countries reporting more cases in males [123].

2.7.1 Epidemiology and pathology

SCC is the most common anal cancer histological type, followed by adenocarcinoma [124–126]. SCC and its precursor lesions, such as anal intraepithelial neoplasia (AIN), are thought to be influenced by HPV infection [127,128]. HPV-induced anal carcinogenicity is influenced by a number of factors, including early sexual activity, previous exposure to HR-HPV genotypes, a history of anogenital injury, and other HPV-related cancers [127–131]. The anal AIN is the precursor lesion for anal cancer. HPV is considered to be responsible for 80%–90% of anal malignancies in general, with HPV-16 being a more common cause (70%) than HPV-18 (10%) [132,133].

2.8 Genital warts (*Condyloma acuminatum*)

Genital warts are also known as anogenital warts or *Condyloma acuminatum*. Genital warts are benign epithelial neoplasm that is associated with HPV-6 or HPV-11 in >90% of cases [134,135] and located in the genito-anal skin and adjacent mucosa. On the scrotum, the shaft of the penis, and around the tip of the penis, genital warts can be detected in men. Genital warts can appear on the vulva, vagina, and cervix in women. Warts can be present in the groin and around the anus in both men and women. Warts around the mouth, tongue, and gums, as well as in the throat, can be caused by oral intercourse [136].

2.8.1 Epidemiology and pathology

Recent studies stated that nearly 19%–44% of genital warts are associated with HPV-6 or HPV-11 and HR or intermediate-risk HPV types [137,138]. The majority of patients are between the ages of 20–40 years [139]. The incidence of *Condyloma acuminatum* is difficult to estimate because anogenital warts aren't a reportable condition. Most genital warts are diagnosed visually, and a confirming biopsy is rarely required. Hyperplastic squamous epithelium containing koilocytes that are squamous epithelial cells with hyperchromatic, acentric nucleus moved by a huge perinuclear vacuole lines these exophytic lesions [140]. Anogenital warts are estimated to affect 1.1 to 1.2 cases per 1000 people in the United States, according to recent estimations [141].

3. Treatment of HPV-associated malignancies

Surgical procedures and vaccinations are commonly used to treat HPV-related cancers. Surgery can be used to treat malignancies ranging from mild to severe. For example, cervical dysplasia is treated by removing abnormal cells from the cervix, which can lower the chance of CC. Cryotherapy, conization, laser therapy, radiation, loop electrosurgical excision process, and other surgical treatments are also available for the treatment of malignancies linked to HPV. The aberrant cells are frozen with liquid nitrogen or CO_2 during cryotherapy. Conization is a procedure that removes aberrant cells from the body. Abnormal cells are destroyed or damaged by laser therapy and radiation (cancerous cells). The loop electrosurgical excision process (LEEP) includes the use of an electrical current to remove aberrant cells.

Vaccines play a major role in the treatment of malignancies driven by HPV. Vaccines for preventing HPV infections and for treatment of HPV-associated malignancies are categorized into prophylactic vaccines and therapeutic vaccines, respectively. Prophylactic vaccines help in preventing a normal person from being infected with HPV and an infected person from being re-infected with HPV. It is not used to treat HPV-related diseases or cancers, but it does help to reduce the liability of people with HPV-related diseases [142,143]. Prophylactic vaccines against HPV's key capsid protein L1 have been produced. According to certain research, prophylactic vaccinations can be considered as an adjuvant that can prevent HPV re-infection following surgical treatment of CIN2 [143,144]. Three vaccines are approved by FDA for preventing the HPV-associated malignancies. Gardasil (4vHPV) is a quadrivalent HPV vaccine approved by the FDA for use against HPVs in 2006. HPV 6, 11, 16, and 18 are all resistant to it. Bivalent Cervarix, a vaccine that provides immunity against HPV 16 and 18, was approved by the FDA in 2007. The FDA approved Gardasil 9 (9vHPV) in 2014, making it more effective against HPV 6, 11, 16, 18, 31, 33, 45, 52, and 58.

4. Conclusion

HPV is the causal agent for one of the largest families of cancers among which CCs and oropharyngeal cancers are the most common cancers in women associated with HR-HPV infection. Lack of HPV screening, misdiagnosis, and low public awareness contribute to the high incidence of HPV-driven malignancies and the mortality and morbidity associated with it. As per studies, HPV-driven cancer incidence trends show a persistent increase in high-income countries such as United Kingdom, Netherlands, Denmark, Australia, Japan, and China, due to geographic differences in sexual behavior associated with HPV infection, low level of screening, and lack or low HPV vaccination programs [36,145].

Regular screening, public awareness, and educational counseling may play a pivotal role in lowering the burden of HPV-driven malignancies as "Prevention is always better than cure". Besides available treatment and prevention methods (such as prophylactic vaccines and some surgical, laser, and radiotherapies), organized cytological screening accompanied with vaccination programs should be the best way to control the HPV-driven malignancies in developing countries with high prevalence and mortality of HPV-associated malignancies.

Acknowledgment

AKK and DK are grateful to the University Grants Commission, New Delhi for Start-Up research grant (M-14-0592 to AKK and M-14-57 to DK), Institute of Eminence (IoE) Seed Grant, Banaras Hindu University (Dev. Scheme No. 6031 to AKK and DK) and Science and Engineering Research Board Start-Up Research Grant (SRG/2021/001732 to AKK) for providing financial assistance. PKT is thankful to Banaras Hindu University for UGC

Research fellowship. The financial support from the Department of Biotechnology (DBT) to the School of Biotechnology, BHU is also acknowledged.

References

[1] C.D. Martel, et al., Worldwide burden of cancer attributable to HPV by site, country and HPV type, Int. J. Cancer 141 (4) (2017) 664–670.
[2] J. Ferlay, et al., Estimating the global cancer incidence and mortality in 2018: GLOBOCAN sources and methods, Int. J. Cancer 144 (8) (2019) 1941–1953.
[3] R.E. Shope, E.W. Hurst, Infectious papillomatosis of rabbits with a note on the histopathology, J. Exp. Med. 58 (5) (1933) 607–624.
[4] Z.M. Zheng, C.C. Baker, Papillomavirus genome structure, expression, and post-transcriptional regulation, Front. Biosci. 11 (2006) 2286–2302.
[5] K. Valentino, C.B. Poronsky, Human Papillomavirus infection and vaccination, J. Pediatr. Nurs. 31 (2) (2016) e155–e166.
[6] C. Chelimo, et al., Risk factors for and prevention of human papillomaviruses (HPV), genital warts and cervical cancer, J. Infect. 66 (3) (2013) 207–217.
[7] U. Sahu, P. Khare, Role of interleukin-17 in human papillomavirus infection and associated malignancies, Microb. Pathog. (2021) 105294.
[8] U. Sahu, et al., Interleukin-17—a multifaceted cytokine in viral infections, J. Cell Physiol. 236 (12) (2021) 8000–8019.
[9] C. Zhou, Z.K. Tuong, I.H. Frazer, Papillomavirus immune evasion strategies target the infected cell and the local immune system, Front. Oncol. 9 (2019) 682.
[10] E.M. de Villiers, Cross-roads in the classification of papillomaviruses, Virology 445 (1–2) (2013) 2–10.
[11] D.M. Parkin, F. Bray, Chapter 2: the burden of HPV-related cancers, Vaccine 24 (Suppl 3) (2006) S3/11–25.
[12] H. zur Hausen, Papillomaviruses and cancer: from basic studies to clinical application, Nat. Rev. Cancer 2 (5) (2002) 342–350.
[13] A.A. McBride, Mechanisms and strategies of papillomavirus replication, Biol. Chem. 398 (8) (2017) 919–927.
[14] F.X. Bosch, et al., Prevalence of human Papillomavirus in cervical cancer: a worldwide perspective, J. Natl. Cancer Inst. 87 (11) (1995) 796–802.
[15] A.B. Moscicki, et al., Updating the natural history of human papillomavirus and anogenital cancers, Vaccine 30 (Suppl 5) (2012) F24–F33.
[16] W.A. Rivoire, et al., Biologia molecular do câncer cervical, Revista Brasileira de Saúde Materno Infantil 6 (4) (2006) 447–451.
[17] P. Bonanni, S. Boccalini, A. Bechini, Efficacy, duration of immunity and cross protection after HPV vaccination: a review of the evidence, Vaccine 27 (SUPPL. 1) (2009) A46–A53.
[18] D. DiMaio, D. Mattoon, Mechanisms of cell transformation by papillomavirus E5 proteins, Oncogene 20 (54) (2001) 7866–7873.
[19] M.L. Gillison, et al., Human papillomavirus and diseases of the upper airway: head and neck cancer and respiratory papillomatosis, Vaccine 30 (SUPPL.5) (2012) F34–F54.
[20] B. Serrano, et al., Human papillomavirus genotype attribution for HPVs 6, 11, 16, 18, 31, 33, 45, 52 and 58 in female anogenital lesions, Eur. J. Cancer 51 (13) (2015) 1732–1741.
[21] A. Biswas, Human papillomavirus (HPV) and cervical cancer, J. Indian Med. Assoc. 98 (2) (2000) 53–55.
[22] M. Grce, M. Mravak-Stipetić, Human papillomavirus–associated diseases, Clin. Dermatol. 32 (2) (2014) 253–258.

[23] N.F. Schlecht, et al., Persistent human papillomavirus infection as a predictor of cervical intraepithelial neoplasia, JAMA 286 (24) (2001) 3106–3114.
[24] E.C. Goodwin, D. DiMaio, Repression of human papillomavirus oncogenes in HeLa cervical carcinoma cells causes the orderly reactivation of dormant tumor suppressor pathways, Proc. Natl. Acad. Sci. U. S. A. 97 (23) (2000) 12513–12518.
[25] E.C. Goodwin, et al., Rapid induction of senescence in human cervical carcinoma cells, Proc. Natl. Acad. Sci. U. S. A. 97 (20) (2000) 10978–10983.
[26] L. Mirabello, et al., The intersection of HPV epidemiology, genomics and mechanistic studies of HPV-mediated carcinogenesis, Viruses 10 (2) (2018).
[27] S.R. Skinner, et al., Progression of HPV infection to detectable cervical lesions or clearance in adult women: analysis of the control arm of the VIVIANE study, Int. J. Cancer 138 (10) (2016) 2428–2438.
[28] M. Stanley, Pathology and epidemiology of HPV infection in females, Gynecol. Oncol. 117 (2 Suppl) (2010) S5–10.
[29] S. Monnier-Benoit, et al., Dynamics of HPV16 DNA load reflect the natural history of cervical HPV-associated lesions, J. Clin. Virol. 35 (3) (2006) 270–277.
[30] M. van Duin, et al., Human papillomavirus 16 load in normal and abnormal cervical scrapes: an indicator of CIN II/III and viral clearance, Int. J. Cancer 98 (4) (2002) 590–595.
[31] P.E. Gravitt, et al., A cohort effect of the sexual revolution may be masking an increase in human papillomavirus detection at menopause in the United States, J Infect Dis 207 (2) (2013) 272–280.
[32] R.P. Insinga, E.J. Dasbach, E.H. Elbasha, Epidemiologic natural history and clinical management of human Papillomavirus (HPV) disease: a critical and systematic review of the literature in the development of an HPV dynamic transmission model, BMC Infect. Dis. 9 (2009) 119.
[33] A.B. Moscicki, et al., Rate of and risks for regression of cervical intraepithelial neoplasia 2 in adolescents and young women, Obstet. Gynecol. 116 (6) (2010) 1373–1380.
[34] E. Weiderpass, et al., Estimates of incidence and mortality of cervical cancer in 2018: a worldwide analysis, Lancet Glob. Health 8 (2) (2020) e191–e203.
[35] M. Plummer, et al., Global burden of cancers attributable to infections in 2012: a synthetic analysis, Lancet Glob. Health 4 (9) (2016) e609–e616.
[36] C. de Martel, et al., Global burden of cancer attributable to infections in 2018: a worldwide incidence analysis, Lancet Glob. Health 8 (2) (2020) e180–e190.
[37] Cervical Cancer — Cancer Stat Facts.
[38] F.T. Cutts, et al., Human papillomavirus and HPV vaccines: a review, Bull. World Health Organ. 85 (9) (2007) 719–726.
[39] A.J. Kombe Kombe, et al., Epidemiology and burden of human papillomavirus and related diseases, molecular pathogenesis, and vaccine evaluation, Front. Public Health 8 (2021).
[40] T.G. Ginindza, et al., Cost analysis of human Papillomavirus-related cervical diseases and genital warts in Swaziland, PLoS One 12 (5) (2017), e0177762.
[41] T.H. Haugen, et al., Trans-activation of an upstream early gene promoter of bovine papilloma virus-1 by a product of the viral E2 gene, EMBO J. 6 (1) (1987) 145–152.
[42] Y. Jing, et al., Phylogeny and polymorphism in the long control regions E6, E7, and L1 of HPV type 56 in women from Southwest China, Mol. Med. Rep. 17 (5) (2018) 7131–7141.
[43] M. Lehoux, C.M. D'Abramo, J. Archambault, Molecular mechanisms of human papillomavirus-induced carcinogenesis, Public Health Genomics 12 (5–6) (2009) 268–280.
[44] T.W. Leung, et al., HPV 16 E2 binding sites 1 and 2 become more methylated than E2 binding site 4 during cervical carcinogenesis, J. Med. Virol. 87 (6) (2015) 1022–1033.

[45] V. Tomaić, Functional roles of E6 and E7 oncoproteins in HPV-induced malignancies at diverse anatomical sites, Cancers 8 (10) (2016).
[46] H.M. Jung, B.L. Phillips, E.K. Chan, miR-375 activates p21 and suppresses telomerase activity by coordinately regulating HPV E6/E7, E6AP, CIP2A, and 14-3-3zeta, Mol. Cancer 13 (2014) 80.
[47] L. Ma, et al., miR-9, a MYC/MYCN-activated microRNA, regulates E-cadherin and cancer metastasis, Nat. Cell Biol. 12 (3) (2010) 247–256.
[48] I.A. Asangani, et al., MicroRNA-21 (miR-21) post-transcriptionally downregulates tumor suppressor Pdcd4 and stimulates invasion, intravasation and metastasis in colorectal cancer, Oncogene 27 (15) (2008) 2128–2136.
[49] S. Bumrungthai, et al., Up-regulation of miR-21 is associated with cervicitis and human Papillomavirus infection in cervical tissues, PLoS One 10 (5) (2015), e0127109.
[50] G. Lao, et al., Mir-155 promotes cervical cancer cell proliferation through suppression of its target gene LKB1, Tumour Biol. 35 (12) (2014) 11933–11938.
[51] M. Turner, E. Vigorito, Regulation of B- and T-cell differentiation by a single microRNA, Biochem. Soc. Trans. 36 (Pt 3) (2008) 531–533.
[52] M. Melar-New, L.A. Laimins, Human papillomaviruses modulate expression of microRNA 203 upon epithelial differentiation to control levels of p63 proteins, J. Virol. 84 (10) (2010) 5212–5221.
[53] I.J. Groves, N. Coleman, Pathogenesis of human papillomavirus-associated mucosal disease, J. Pathol. 235 (4) (2015) 527–538.
[54] L. Bucchi, et al., Estimating the impact of an organised screening programme on cervical cancer incidence: a 26-year study from northern Italy, Int. J. Cancer 144 (5) (2019) 1017–1026.
[55] A. Zucchetto, et al., Screening patterns within organized programs and survival of Italian women with invasive cervical cancer, Prev. Med. 57 (3) (2013) 220–226.
[56] E. Altobelli, et al., HPV-vaccination and cancer cervical screening in 53 WHO European countries: an update on prevention programs according to income level, Cancer Med. 8 (5) (2019) 2524–2534.
[57] M. Ansari, et al., Smear patterns and spectrum of premalignant and malignant cervical epithelial lesions in postmenopausal Indian women: a hospital-based study, Diagn. Cytopathol. 40 (11) (2012) 976–983.
[58] P.L. Sachan, et al., A study on cervical Cancer screening using pap smear test and clinical correlation, Asia Pac. J. Oncol. Nurs. 5 (3) (2018) 337–341.
[59] M. Arbyn, et al., EUROGIN 2011 roadmap on prevention and treatment of HPV-related disease, Int. J. Cancer 131 (9) (2012) 1969–1982.
[60] A.K. Chaturvedi, et al., Human papillomavirus and rising oropharyngeal cancer incidence in the United States, J. Clin. Oncol. 29 (32) (2011) 4294–4301.
[61] M. van Seters, et al., Treatment of vulvar intraepithelial neoplasia with topical imiquimod, N. Engl. J. Med. 358 (14) (2008) 1465–1473.
[62] A. Argiris, et al., Head and neck cancer, Lancet 371 (9625) (2008) 1695–1709.
[63] M.L. Gillison, et al., Evidence for a causal association between human papillomavirus and a subset of head and neck cancers, J. Natl. Cancer Inst. 92 (9) (2000) 709–720.
[64] E.M. Rettig, G. D'Souza, Epidemiology of head and neck cancer, Surg. Oncol. Clin. N. Am. 24 (3) (2015) 379–396.
[65] J. Mork, et al., Human papillomavirus infection as a risk factor for squamous-cell carcinoma of the head and neck, N. Engl. J. Med. 344 (15) (2001) 1125–1131.
[66] E.M. Smith, et al., Human papillomavirus seropositivity and risks of head and neck cancer, Int. J. Cancer 120 (4) (2007) 825–832.
[67] F. Bray, et al., Global cancer statistics 2018: GLOBOCAN estimates of incidence and mortality worldwide for 36 cancers in 185 countries, CA Cancer J. Clin. 68 (6) (2018) 394–424.

[68] M. Hashibe, et al., Alcohol drinking in never users of tobacco, cigarette smoking in never drinkers, and the risk of head and neck cancer: pooled analysis in the international head and neck Cancer epidemiology consortium, J. Natl. Cancer Inst. 99 (10) (2007) 777–789.
[69] H. Jiang, et al., Can public health policies on alcohol and tobacco reduce a cancer epidemic? Australia's experience, BMC Med. 17 (1) (2019) 213.
[70] H. Mehanna, et al., Prevalence of human papillomavirus in oropharyngeal and nonoropharyngeal head and neck cancer—systematic review and meta-analysis of trends by time and region, Head Neck 35 (5) (2013) 747–755.
[71] S. de Sanjose, et al., Burden of human papillomavirus (HPV)-related cancers attributable to HPVs 6/11/16/18/31/33/45/52 and 58, JNCI Cancer Spectr. 2 (4) (2018) pky045.
[72] B.A. Mahal, et al., Incidence and demographic burden of HPV-associated oropharyngeal head and neck cancers in the United States, Cancer Epidemiol. Biomarkers Prev. 28 (10) (2019) 1660–1667.
[73] V. Senkomago, et al., Human Papillomavirus-attributable cancers - United States, 2012-2016, MMWR Morb. Mortal. Wkly Rep. 68 (33) (2019) 724–728.
[74] J. Califano, et al., Genetic progression model for head and neck cancer: implications for field cancerization, Cancer Res. 56 (11) (1996) 2488–2492.
[75] Cancer Genome Atlas Network, Comprehensive genomic characterization of head and neck squamous cell carcinomas, Nature 517 (7536) (2015) 576–582.
[76] A.R. Kreimer, A.K. Chaturvedi, HPV-associated oropharyngeal cancers—are they preventable? Cancer Prev. Res. (Phila.) 4 (9) (2011) 1346–1349.
[77] D.J.G.O. Hunter, Carcinoma of the vulva: a review of 361 patients, Gynecol. Oncol. 3 (2) (1975) 117–123.
[78] R. Siegel, D. Naishadham, A. Jemal, Cancer statistics, 2013, CA Cancer J. Clin. 63 (1) (2013) 11–30.
[79] T. Iversen, S. Tretli, Intraepithelial and invasive squamous cell neoplasia of the vulva: trends in incidence, recurrence, and survival rate in Norway, Obstet. Gynecol. 91 (6) (1998) 969–972.
[80] S. Pilotti, et al., Study of multiple human papillomavirus-related lesions of the lower female genital tract by in situ hybridization, Hum. Pathol. 20 (2) (1989) 118–123.
[81] J. Scurry, E.J. Wilkinson, Review of terminology of precursors of vulvar squamous cell carcinoma, J. Low. Genit. Tract Dis. 10 (3) (2006) 161–169.
[82] F. Bonvicini, et al., Presence and type of oncogenic human papillomavirus in classic and in differentiated vulvar intraepithelial neoplasia and keratinizing vulvar squamous cell carcinoma, J. Med. Virol. 77 (1) (2005) 102–106.
[83] R. Sankaranarayanan, J. Ferlay, Worldwide burden of gynaecological cancer: the size of the problem, Best Pract. Res. Clin. Obstet. Gynaecol. 20 (2) (2006) 207–225.
[84] C.P. Crum, Carcinoma of the vulva: epidemiology and pathogenesis, Obstet. Gynecol. 79 (3) (1992) 448–454.
[85] L.N. Hoang, et al., Squamous precursor lesions of the vulva: current classification and diagnostic challenges, Pathology 48 (4) (2016) 291–302.
[86] M.C.G. Bleeker, et al., Penile cancer: epidemiology, pathogenesis and prevention, World J. Urol. 27 (2) (2009) 141–150.
[87] B.Y. Hernandez, et al., Burden of invasive squamous cell carcinoma of the penis in the United States, 1998-2003, Cancer 113 (10 SUPPL) (2008) 2883–2891.
[88] P. Kirrander, et al., Swedish National Penile Cancer Register: incidence, tumour characteristics, management and survival, BJU Int. 117 (2) (2016) 287–292.
[89] T.B. Olesen, et al., Prevalence of human papillomavirus DNA and p16INK4a in penile cancer and penile intraepithelial neoplasia: a systematic review and meta-analysis, Lancet Oncol. 20 (1) (2019) 145–158.

[90] B. Serrano, et al., Epidemiology and burden of HPV-related disease, Best Pract. Res. Clin. Obstet. Gynaecol. 47 (2018) 14–26.
[91] J. Dillner, et al., Etiology of squamous cell carcinoma of the penis, Scand. J. Urol. Nephrol. Suppl. 34 (205) (2000) 189–193.
[92] M.A. Rubin, et al., Detection and typing of human Papillomavirus DNA in penile carcinoma: evidence for multiple independent pathways of penile carcinogenesis, Am. J. Pathol. 159 (4) (2001) 1211–1218.
[93] H.R. Wabinga, et al., Trends in cancer incidence in Kyadondo County, Uganda, 1960–1997, Br. J. Cancer 82 (9) (2000) 1585–1592.
[94] P.J.F. Snijders, et al., HPV-mediated cervical carcinogenesis: concepts and clinical implications, J. Pathol. 208 (2) (2006) 152–164.
[95] A. Flaherty, et al., Implications for human papillomavirus in penile cancer, Urol. Oncol. 32 (1) (2014). p. 53 e1-8.
[96] P.E. Spiess, et al., Pathophysiological basis of human papillomavirus in penile cancer: key to prevention and delivery of more effective therapies, CA Cancer J. Clin. 66 (6) (2016) 481–495.
[97] C. Annunziata, et al., Characterization of the human papillomavirus (HPV) integration sites into genital cancers, Pathol. Oncol. Res. 18 (4) (2012) 803–808.
[98] O. Aynaud, M. Ionesco, R.J.C. Barrasso, Penile intraepithelial neoplasia. Specific clinical features correlate with histologic and virologic findings, Cancer 74 (6) (1994) 1762–1767.
[99] A. Antonsson, et al., High-risk human papillomavirus in esophageal squamous cell carcinoma, Cancer Epidemiol. Biomarkers Prev. 19 (8) (2010) 2080–2087.
[100] S. Herbster, et al., HPV infection in Brazilian patients with esophageal squamous cell carcinoma: interpopulational differences, lack of correlation with surrogate markers and clinicopathological parameters, Cancer Lett. 326 (1) (2012) 52–58.
[101] X. Li, et al., Systematic review with meta-analysis: the association between human papillomavirus infection and oesophageal cancer, Aliment. Pharmacol. Ther. 39 (3) (2014) 270–281.
[102] S.K. Zhang, et al., The association between human papillomavirus 16 and esophageal cancer in Chinese population: a meta-analysis, BMC Cancer 15 (2015) 1096.
[103] M. Hongo, Y. Nagasaki, T. Shoji, Epidemiology of esophageal cancer: orient to occident. Effects of chronology, geography and ethnicity, J. Gastroenterol. Hepatol. 24 (5) (2009) 729–735.
[104] S.S. Devesa, W.J. Blot, J.F. Fraumeni Jr., Changing patterns in the incidence of esophageal and gastric carcinoma in the United States, Cancer 83 (10) (1998) 2049–2053.
[105] H. Pohl, B. Sirovich, H.G. Welch, Esophageal adenocarcinoma incidence: are we reaching the peak? Cancer Epidemiol. Biomarkers Prev. 19 (6) (2010) 1468–1470.
[106] H. Pohl, H.G. Welch, The role of overdiagnosis and reclassification in the marked increase of esophageal adenocarcinoma incidence, J. Natl. Cancer Inst. 97 (2) (2005) 142–146.
[107] U. Wieland, A. Kreuter, H. Pfister, Human Papillomavirus and Immunosuppression, Curr. Probl. Dermatol. (Switz.) 45 (2014) 154–165.
[108] F. Chang, et al., Screening for human papillomavirus infections in esophageal squamous cell carcinomas by in situ hybridization, Cancer 72 (9) (1993) 2525–2530.
[109] T. Matsha, et al., Human papillomavirus associated with oesophageal cancer, J. Clin. Pathol. 55 (8) (2002) 587–590.
[110] N.F. Hacker, et al., Outcome and patterns of recurrence for International Federation of Gynecology and Obstetrics (FIGO) stages i and II squamous cell vulvar cancer, Obstet. Gynecol. 113 (4) (2009) 895–901.
[111] A. Kubo, D.A. Corley, Marked multi-ethnic variation of esophageal and gastric cardia carcinomas within the United States, Am. J. Gastroenterol. 99 (4) (2004) 582–588.

[112] R.W. Jones, D.M. Rowan, A.W. Stewart, Vulvar intraepithelial neoplasia: aspects of the natural history and outcome in 405 women, Obstet. Gynecol. 106 (6) (2005) 1319–1326.
[113] U. Beller, et al., Carcinoma of the vulva, J. Epidemiol. Biostat. 6 (1) (2001) 153–174.
[114] G.S. Steele, et al., The National Cancer Data Base report on patterns of care for testicular carcinoma, 1985-1996, Cancer 86 (10) (1999) 2171–2183.
[115] T. Bjorge, et al., Prospective seroepidemiological study of role of human papillomavirus in non-cervical anogenital cancers, BMJ 315 (7109) (1997) 646–649.
[116] J.J. Carter, et al., Human papillomavirus 16 and 18 L1 serology compared across anogenital cancer sites, Cancer Res. 61 (5) (2001) 1934–1940.
[117] J.R. Daling, et al., A population-based study of squamous cell vaginal cancer: HPV and cofactors, Gynecol. Oncol. 84 (2) (2002) 263–270.
[118] M. Hampl, et al., Effect of human papillomavirus vaccines on vulvar, vaginal, and anal intraepithelial lesions and vulvar cancer, Obstet. Gynecol. 108 (6) (2006) 1361–1368.
[119] J.L. Benedet, B.H. Sanders, Carcinoma in situ of the vagina, Am. J. Obstet. Gynecol. 148 (5) (1984) 695–700.
[120] M. Frisch, J.H. Olsen, M. Melbye, Malignancies that occur before and after anal Cancer: clues to their etiology, Am. J. Epidemiol. 140 (1) (1994) 12–19.
[121] S.K. Hussain, J. Sundquist, K. Hemminki, Familial clustering of cancer at human papillomavirus-associated sites according to the Swedish family-Cancer database, Int. J. Cancer 122 (8) (2008) 1873–1878.
[122] S. Hartwig, et al., Estimation of the epidemiological burden of human papillomavirus-related cancers and non-malignant diseases in men in Europe: a review, BMC Cancer 12 (2012).
[123] D. Schottenfeld, J.F. Fraumeni, Cancer epidemiology and prevention, Cancer Epidemiol. Prev. 9780195149 (2009) 1–1410.
[124] L. Abramowitz, et al., Human papillomavirus genotype distribution in anal cancer in France: the EDiTHV study, Int. J. Cancer 129 (2) (2011) 433–439.
[125] L. Alemany, et al., Human papillomavirus DNA prevalence and type distribution in anal carcinomas worldwide, Int. J. Cancer 136 (1) (2015) 98–107.
[126] P.M. Hoff, R. Coudry, C.M.V. Moniz, Pathology of anal cancer, Surg. Oncol. Clin. N. Am. 26 (1) (2017) 57–71.
[127] C.C. Chen, Y.Y. Chou, Predictive value of the anal cytology for detecting anal intraepithelial neoplasia or worse: a systematic review and meta-analysis, Diagn. Cytopathol. 47 (4) (2019) 307–314.
[128] M. Morton, N. Melnitchouk, R. Bleday, Squamous cell carcinoma of the anal canal, Curr. Probl. Cancer 42 (5) (2018) 486–492.
[129] R.D. Cranston, et al., Prevalence and determinants of anal human papillomavirus infection in men who have sex with men and transgender women, Int J STD AIDS 30 (2) (2019) 154–162.
[130] A.-B. Moscicki, et al., The role of sexual behavior and human papillomavirus persistence in predicting repeated infections with new human papillomavirus types, Cancer Epidemiol. Biomarkers Prev. 19 (8) (2010) 2055–2065.
[131] R.A. Nelson, et al., Changing patterns of anal canal carcinoma in the United States, J. Clin. Oncol. 31 (12) (2013) 1569–1575.
[132] B.E. Hoots, et al., Human papillomavirus type distribution in anal cancer and anal intraepithelial lesions, Int. J. Cancer 124 (10) (2009) 2375–2383.
[133] H.D.Vuyst, et al., Prevalence and type distribution of human papillomavirus in carcinoma and intraepithelial neoplasia of the vulva, vagina and anus: a meta-analysis, Int. J. Cancer 124 (7) (2009) 1626–1636.
[134] L. Gissmann, et al., Human papillomavirus types 6 and 11 DNA sequences in genital and laryngeal papillomas and in some cervical cancers, Proc. Natl. Acad. Sci. 80 (2) (1983) 560–563.

[135] G. Gross, et al., Papillomavirus infection of the Anogenital region: correlation between histology, clinical picture, and virus type. Proposal of a New nomenclature, J. Investig. Dermatol. 85 (2) (1985) 147–152.
[136] D. Grennan, Genital warts, JAMA 321 (5) (2019) 520.
[137] D.R. Brown, et al., Detection of multiple human papillomavirus types in condylomata acuminata lesions from otherwise healthy and immunosuppressed patients, J. Clin. Microbiol. 37 (10) (1999) 3316–3322.
[138] K. Hong, et al., Isolation and characterization of human papillomavirus type 6-specific T cells infiltrating genital warts, J. Virol. 71 (9) (1997) 6427–6432.
[139] A.B. Fleischer Jr., et al., Condylomata acuminata (genital warts): patient demographics and treating physicians, Sex. Transm. Dis. 28 (11) (2001) 643–647.
[140] A.K. Leung, et al., Penile warts: an update on their evaluation and management, Drugs Context 7 (2018) 212563.
[141] I.U. Park, C. Introcaso, E.F. Dunne, Human Papillomavirus and genital warts: a review of the evidence for the 2015 centers for disease control and prevention sexually transmitted diseases treatment guidelines, Clin. Infect. Dis. 61 (Suppl 8) (2015) S849–S855.
[142] B. Ma, et al., Emerging human papillomavirus vaccines, Expert Opin. Emerg. Drugs 17 (4) (2012) 469–492.
[143] D.M. Harper, K.B. Williams, Prophylactic HPV vaccines: current knowledge of impact on gynecologic premalignancies, Discov. Med. 10 (50) (2010) 7–17.
[144] N. Kash, et al., Safety and efficacy data on vaccines and immunization to human Papillomavirus, J. Clin. Med. 4 (4) (2015) 614–633.
[145] J. Wu, et al., Worldwide trend in human papillomavirus–attributable cancer incidence rates between 1990 and 2012 and Bayesian projection to 2030, Cancer 127 (17) (2021) 3172–3182.

CHAPTER 4

Immunological aspects of HPV/HIV co-infection and risk of gynecological diseases

Amrita Srivastava[a], Priyanka Namdev[a,b], and Rajaneesh Anupam[a,#]
[a]Department of Biotechnology, Dr. Harisingh Gour Vishwavidyalaya (A Central University), Sagar, Madhya Pradesh, India
[b]Centers for Disease Control and Prevention, Atlanta, GA, United States

1. Introduction

The human papillomavirus (HPV) belongs to family *Papillomaviridae*. It is a small, nonenveloped virus with double-stranded DNA genome and approximately 55 nm in diameter. Out of more than 150 types of HPVs, around 40 are capable of causing genital tract infection in humans. The genital HPVs are further grouped as low-risk types (LR-HPV) as they are nononcogenic and high-risk types (HR-HPV) due to their oncogenic property. Although the LR-HPVs are not associated with cancer, they cause genital warts and benign/low-grade cervical cellular changes. The HR-HPVs specifically types 16 and 18 are responsible for high-grade cervical cellular changes and cervical cancer (CC). Besides, a few other HR-HPVs are involved in cancers of the vulva, vagina, anus, penis, and oropharynx [1]. HPV infection is one of the highly prevalent viral infections in the world. The genital HPV infection rate in both sexes is similar at 3.5%–45% in men and 2%–44% in women [2,3].

The molecular structure of HPV genome shows the presence of eight open reading frames, six are early genes (E1, E2, E4, E5, E6, and E7) coding early proteins and two are late genes (L1 and L2) coding late proteins [4]. After the entry of viral particle (viral DNA and L1 and L2 capsid proteins, which form icosahedral capsid) into the basal lamina cell which delivers viral genome into the nucleus, the life cycle of the virus has three phases. Initial phase involves maintaining a constant lower copy number of viral episome in the cell through the action E1 and E2 proteins along with cellular proteins. In the second phase, E6 and E7 proteins promote mid/upper epithelial

[#] Current affiliation: Department of Biochemistry and Molecular Biology, Pondicherry University, Pondicherry, India.

Immunopathology, Diagnosis, and Treatment of HPV-induced Malignancies
https://doi.org/10.1016/B978-0-323-91797-1.00004-1

cell proliferation which leads to huge increase in the copy number of the viral genome. In the final phase, the viral gene expression switches to the production of L1 and L2 proteins that pack the viral genome to form and release new viral particles [1]. E5, E6, and E7 proteins are implicated in viral immune evasion. The expression of histocompatibility leukocyte antigen (HLA) class I is downregulated by E5 protein to escape detection by $CD8^+$ T cells [5–7]. Oncogenic E6 and E7 proteins are vital for transformation of HPV-infected cells. E6 binds to p53 and PDZ-containing proteins to promote their ubiquitin-mediated degradation resulting in cell cycle progression. The telomerase expression is also increased by E6 to support cell proliferation. Similarly, E7 binds to retinoblastoma (Rb) tumor suppressor family proteins to inhibit their function and degrade them to facilitate S phase entry which is required for viral replication. Ultimately, transformation of HPV infected cells occurs due the collective action of E6 and E7 that dysregulate the cell cycle check points to promote cell proliferation which leads to accumulation of mutations. The integration of HPV genome is an important event for transformation as it disrupts E2 region resulting loss of E6 and E7 regulation by E2. [4]. The E6 and E7 expression is the primary incidence in HR-HPV infection and cervical intraepithelial neoplasia (CIN) development. E7 lowers Toll-like receptor (TLR)-9 expression, which activates antigen-presenting cells by acting as a viral DNA sensor [8]. Transmission of genital HPV takes place primarily through sexual intercourse with an infected individual. Nonsexual routes of transmission includes vertical transmission from mother to infant [9]. The primary infection site of HPV is basal cell layer of stratified squamous epithelial cells of cervix skin. The virus gets an entry into the basement membrane of cervix skin when it gets exposed by micro-abrasion. Viral protein-driven proliferation of the epithelial layer causes hyperplasia and dysplasia leading to carcinoma. The percentage of dysplastic cells of cervical epithelium determines the grade of the dysplasia. Cervical intraepithelial neoplasia (CIN)-1, considered low-grade, covers the lower 1/3 part or less of the epithelium, while the more important CIN-2 and CIN-3 (high-grade) grows to cover the whole depth of the epithelium [10]. The three main viral proteins E5, E6, and E7 are key game changers in the infection by evading immune system [11]. Cervical carcinomas are detected by Papanicolaou test, generally called as Pap smear test which screens cervical cells for cytological abnormalities [12]. The HPV infections can be prevented by abstaining from sexual contact with multiple partners, by regular condom use and HPV vaccination available to effectively manage HPV infections and associated diseases [13].

The human immunodeficiency virus (HIV), belongs to the *Retroviridae* family, is an enveloped RNA virus that contains pair of single stranded sense RNA. It is a spherical virus with its envelope embedded with viral glycoproteins encoded by the *env* gene, which binds to CD4 receptor on T cells to gain entry into the cell. The nucleocapsid protein that binds to viral genomic RNA, the capsid core holding the viral RNA-nucleocapsid complex, and the matrix which lines this core are encoded by *gag* gene. Upon entry, the viral RNA is converted into DNA and integrated into host genome by *pro* gene-encoded viral reverse transcriptase and integrase (that virus brings with itself). The integrated provirus uses host machinery to replicate and assemble new viral particles that get released through budding. HIV genome also has six additional genes (*nef, rev, tat, vif, vpr,* and *vpu/vpx*) that encode six regulatory and accessory proteins which play important roles in HIV life cycle and pathogenesis [14].

The later stages of HIV infection lead to acquired immunodeficiency syndrome (AIDS), which is characterized by opportunistic infections and tumors that lead to death. HIV-1 and HIV-2 are the two main HIV types, where HIV-1 is more globally expanded and virulent. As per UNAIDS in 2020, around 37.6 million population is infected with HIV/AIDS globally [15]. HIV is transmitted by the exchange of certain types of bodily fluids, for example, blood, semen, breast milk, and vaginal fluids. HIV is mostly transmitted through sex. A HIV^+ mother can also transmit the infection to the child during pregnancy, delivery, or through breastfeeding [16].

As HPV infection causes CC, it is natural that the association between CC and HIV infection is looked through connection between HPV and HIV infections. HPV infection can be assisted by HIV co-infection. It is a well-established fact that women population infected with HIV are immunocompromised and highly vulnerable to develop CC by the virtue of high risk of HPV infection and its persistence [17]. Wide range of HPV genotypes and multiple concurrent infections are remarkably observed in HIV^+ women. This finding is very well reflected by high prevalence rate, i.e., 73.5% of HPV in HIV^+ women [18]. The prevalence of HPV in HIV^+ women was in range of 54%–73% in United States [19,20]. Although European countries have shown slightly lower prevalence of HPV in HIV^+ women which was about 44% [21], occurrence of HPV infection is lower in HIV^- population as compared with HIV^+ population. In HIV^- women with median age of 25 years, the prevalence of any type of HPV was 47.0% which is much lower than its HIV^+ counterparts which was 72.2% [22]. The HIV^+ women are highly vulnerable and susceptible to HPV infection due to poor cell-mediated immunity that fails to knock out the infections. Such kind of immune suppression will also result in resurgence of dormant HPV infections that ends up in enhanced incidence, prevalence, and persistence of HPV infection [23]. Most prominently, the HIV$^+$ patients are highly susceptible to HPV16 and HPV56 genotypes co-infection [17]. The possible reason behind HPV co-infection is due to retarded immune system leading to immune deficiency, and this creates high susceptibility toward HPV56 infection in HIV-infected individuals.

Reports have shown that the high rate of HPV infection in HIV$^+$ population and as expected the occurrence of high-grade squamous intraepithelial lesions (HSILs) are also common in this group. HR-HPV is the most predominant in establishing cytological abnormalities among the HIV-HPV co-infected female subjects. Routine screening of these infections should be embraced to detect and prevent the cancer. In conclusion, it is clear that the co-infection with HIV and interplay between two viruses is pivotal to alter the variety of HPV types and in the development of the associated diseases and cancer.

In this chapter, we discuss various mechanisms through which HPV and HIV modulate immune system to not only facilitate their own infection but also create an environment that is conducive for the infection by the other virus [24]. The facilitation of HPV infection in HIV$^+$ women leading to gynecological cancer and disease is also discussed.

2. Immunological aspects of HPV infection

The incubation period of HPV infection is approximately 10 years, and the host immune response is very crucial in the advancement or deterioration of HR-HPV infection into uterine cervix. An efficient immune response leads to the viral clearance, while an inefficient immune response usually facilitates the viral pathological processes and progression to high-grade neoplasia [25]. Infection with HR-HPV involves cascade of events, which are necessary for CIN development and carcinogenesis. The HR-HPV infection triggers the migration of various immune cells to the dermis. In the squamous epidermis macrophages, keratinocytes (KCs), langerhans cells (LC), T lymphocytes, dendritic cells (DCs), natural killer cells (NKCs), and B lymphocytes play crucial roles in immune response to HPV infection. HR-HPV infection makes immune system highly tolerant to the infection creating a susceptible microenvironment to facilitate the development and progression of CIN [26].

HR-HPV modulates host immune response through various mechanisms to aid its persistent infection and facilitate development of CC. HR-HPV remains dormant for years; during this period, its replication and assembly does not lead to host cell death [27]. E6 and E7 oncoproteins specifically interact with interferon (IFN) regulatory factors (IRF)-1 and IRF-3 to interfere with IFN signaling pathways to retard the IFN synthesis [28,29]. The HR-HPV downregulates the expression of pro-inflammatory cytokine tumor necrosis factor (TNF)-α whereas antiinflammatory cytokine IL-10 is upregulated to avoid immune clearance [30]. The infiltration of T regulatory (Treg) cells is stimulated by HR-HPV infection, which secretes IL-10 or transforming growth factor β (TGF-β) [26]. HPV decreases the expression of MHC class I on the infected cells rendering nonfunctional CTLs [31]. Thereafter, an accumulation of ineffective $CD4^+$ and $CD8^+$ T lymphocytes is seen in second or third stage of CINs [32]. Additionally, KCs present in epidermis are important cells for cytokine secretion. The suppression of these important innate immunity cells is a major reason for infection of host cells with HR-HPV and development of preneoplastic lesions [33].

HR-HPV starts to alter the host innate immunity as soon as it infects mucosal epithelium. The number of LCs that reside in the epidermis and in the transformation zone is significantly decreased. The migration of precursors of LCs to the epidermis is determined by chemokine C-C motif ligand 20 (CCL20). E7 protein binds to CCAAT/enhancer-binding protein β, which is the transcription factor of CCL20, to inhibit CCL20 expression to hinder the migration of precursors of LCs to the epidermis [34]. TLRs in the endolysosomal compartments are the pattern recognition receptors specialized in sensing the virus infections and triggering the downstream innate immune responses. The downregulation of TLR9 expression (nucleic acid-specific receptors) by E6 and E7 proteins of HR-HPV 18 is the primary strategy of HPV to escape immune surveillance [35]. Contrary to TLR9, the pathways of TLR3/5/8 get turned on in HR-HPV-infected KCs. Studies have demonstrated that the higher TLR8 levels in CC cells are due to an increase in expression of Bcl-2 and vascular endothelial growth factor (VEGF) [36]. Another study demonstrated that there is noticeable decrease in TLR4 expression in HR-HPV positive cells, and it is linked to the type of virus and the histopathological grade. HPV16 and CC show higher TLR4 expression compared to HPV18 and CIN. Additionally, the opposition of the HPV-infected cells to apoptosis is due to TLR4 overexpression [36]. Cancer cell migration, proliferation, and angiogenesis are promoted by M2 macrophages and can restrict the immune defense mechanisms [37]. During progression of cervical lesions, the M2 macrophages infiltrate and trigger the advancement of HPV16-associated tumors by inhibiting antitumor T-cell response, causing tumor growth [38]. Additionally, NKC activation is being hampered in HR-HPV infections. In the early stages of infection, NKCs population identifies dominant and low-grade lesions, but in the advanced stages of infection, NKC-activating receptors such as NKG2D, NKp30, NKp45, NKp46, and NKp80 levels are predominantly reduced in HPV16 CC [39]. The dramatic change in the host immune response to HR-HPV is due to compromised adaptive and innate immunity. It is important for the advancement of HR-HPV infection to CC. The immune cell (including T cells) profile changes as the CIN and carcinogenesis progresses through different stages indicating that the immune response reflects the various stages of disease progression.

3. The impact of immunological aspects of HPV on HIV infection

HPV and HIV each display interactions with host-favoring infection of each other at the cellular level. The acquisition of HIV is favored by HPV infection in men and women, and burden of dysplasia and cancer due to HPV is high in HIV$^+$ individuals due to immune suppression. Although HPV infection is concomitant with higher rates of HIV acquisition, still the underlying biological mechanisms between the two are unexplored and an open field to study [40]. A persistent oncogenic HPV and HIV co-infection contributes to rapid progression or aggressive cervical carcinogenesis when compared to HIV$^-$ women with a single, multiple, or no HPV infections. The vaginal cytokine profiles of HPV$^+$ females overlap with cytokines associated with high risk of HIV infection but not in the HPV$^-$ subjects [40]. In another study, authors have compared the level of IL-8 in cervicovaginal samples of HIV$^+$ subjects with and without genital tract dysplasia or histologic or cytologic indication of HPV infection, and reported that the IL-8 levels were significantly high in HIV-HPV co-infected women than in the women without HPV. However, other cytokines, like macrophage inflammatory protein 1a (MIP-1a) and regulated-on-activation normal T-expressed and secreted factor (RANTES) which are reported to suppress HIV replication, promoting T helper cells (Th)-1 immune responses against pathogen, were not so different in these groups [41]. The production of

IL-8, thus, could be participating in preventing the immune response or helping virus production from the epithelial cells [42].

The host immune response to HPV is mediated by T-lymphocytes [43], and this response may increase HIV risk since T lymphocytes are prime target cells for HIV. These cells were upregulated in HPV-infected cervical tissues. Further, increased T-lymphocyte influx in case of nonpersistence HPV infections is linked with higher risk of HIV attainment when compared to the persistent HPV infection with lower T-lymphocyte influx [9,44]. The cytokine IL-1β, a key player in immunity against microbes and autoimmune inflammation, acts directly on T lymphocytes expansion and differentiation in vivo to induce protective immunity. Elevated levels of IL-1β, which activates HIV promoter [45], are noticed in HPV-associated abnormal cervical cytology [46].

Defensins (antimicrobial peptide) and thrombospondins, which are considered as anti-HIV proteins, were reported in low amounts in precancerous cervical lesions [47]; however, there was no association between cytological abnormalities and HIV acquisition [9]. Similarly, in HPV-infected cervical squamocolumnar junctions, expression of defensin was reportedly low when compared with vulvar epithelia which supports that there is inadequate recognition of viral antigens by innate immune system for presentation to the adaptive immune system [48]. Hypermethylation of thrombospondin promoter region has been reported in HPV-associated genital cancers in males [49]. These findings suggest the key role of HPV infections in epigenetic changes in host and poor prognosis of infection and cancer progression.

It is biologically anticipated that prevalent HPV may enhance the risk of HIV infection. The permeability of the genital lining to HIV might be increased through epigenetic modifications. E7 protein downregulates epithelial adhesion molecule E-cadherin which also hampers the movement of LCs through the stratified epithelium which captures and presents the antigens such as HIV. Furthermore, HR-HPV infection has been described to decrease the frequency, reduces tissue density, and alters the LC's morphology in the female genital tract [50]. However, HIV first encounters and infects LCs which transmit it to T cells. The consequence of low LCs in the context of HPV infection may be favoring to HIV infection [51,52]. Restricted movement of LCs due to the reduced level of E-cadherin helps virus to hide the infection and compromise the immune system resulting in more HPV acquisition, prevalence, persistence, and disease progression, and could simultaneously intensify the susceptibility toward HIV infection.

HPV and HIV infections create brutal circle that may contribute to higher rates of these infections in some parts of the world. Highly active antiretroviral therapy (HAART) could reduce HPV infections and associated diseases in HIV-infected individuals. However, it is only helpful after several years of treatment which results in lower HIV viremia and considerable immune reconstitution. However, the therapy was not successful in reducing the risk of developing invasive CC [53,54]. Further, in the ART era, anal and cervical cancer prevalence has not reduced. Although numerous reports associate rapid and invasive CC progression with HIV infection, none of these reports could establish a direct relationship between HIV/HPV co-infection and rate of CC progression and severity of disease invasiveness [55].

HIV-infected people are more prone to all types of oral HPV. The molecular interaction between the HIV and HPV and the tolerant immune atmosphere created by HIV infection results in persistence of HPV during HIV/HPV co-infections. The mathematical model by Verma et al. [56] shows that the HPV infection continues to persist in the short run in a co-infected patient undergoing ART. Besides immunological factors, interactions at the molecular level between HIV genes *tat, rev,* and *vpr* and HPV are also reported. Multiple studies indicate an upregulation of HPV oncogenic genes (E6 and E7) expression by Tat [57,58]. The shedding of HPV is increased by HIV Tat protein suggesting that HPV pathobiology is influenced by HIV infection [56]. A mathematical model predicted that antiretroviral therapy would be effective in decreasing HPV infections only when the prolonged treatment restores the normal immune response and $CD4^+$ T cell number [56,59].

4. HPV-HIV co-infections and gynecological disorders

HPV and HIV are the most prevalent sexually transmitted infections with extensive public health impact worldwide. In women, 99.9% of CC cases are associated with HR-HPV [60]. Various epidemiological studies have reported the role of sexually transferred pathogens in the development and pathogenesis of CC; still, specific role of these pathogens in the time course of HPV infection is limited. Most common types of lower genital tract disease caused by HPV in HIV^+ women are as follows: cervical low-grade and high-grade squamous intraepithelial lesion and intraepithelial neoplasia [19], anogenital warts, vulvar and vaginal intraepithelial neoplasia [60–62], and cervical and anal cancers [63]. Table 1 shows HIV-HPV co-infection-related

Table 1 Common types of lower genital tract diseases in HIV^+ women caused by HPV infection and their risk factors.

Diseases	Risk factors	References
CIN	Persistent HPV infection, immunosuppression, sex at early age, high parity, history of abnormal Papanicolaou (Pap) test results, low socioeconomic class, and smoking	[19]
Genital warts	Younger age, high number of sexual partners lifetime, smoking, previous history of genital chlamydial infection, gonorrhea, genital herpes, or trichomoniasis	[62,65,66]
Vulvar intraepithelial neoplasia	Age > 50 years, immunosuppression, metasynchronous vaginal or intraepithelial neoplasia, positive excision margins, and adjacent lichen sclerosis or HPV infection	[62,67]
Anal intraepithelial neoplasia	History of anal sex, immunosuppression, increased lifetime number of sexual partners, genital warts, smoking, receptive anal intercourse	[68,69]
Vaginal intraepithelial neoplasia	History of CIN or CC	[62]
Anal cancer	Males have sex with males	[68,70]
CC	Having HPV^+ male sexual partners, having partners who are sex workers, long-term smoking, use of oral contraceptive and high parity	[64]

gynecological disorders in women and associated risk factors. Most of these diseases are developed as a result of compromised immune response and associated risk factors such as smoking habits, multiple sexual partners or having HPV-infected partner or male sexual partners (HPV-infected), long-term oral contraceptive use, and high parity [64].

4.1 CC pathogenesis in HIV/HPV co-infection

The pathogenesis of CC involves various steps such as integration of HPV episomal DNA into the host genome after which HPV oncoproteins E6 and E7 help in the accumulation of genetic alterations in the host genome [71,72]. Furthermore, there is an interaction between HPV and HIV co-infection, resulting in an increased possibility of HPV-associated morbidity and CC mortality among HIV^+ women [73]. Women with low $CD4^+$ T cell number and high HIV viral load had elevated risk of HPV infection and decreased HPV clearance, but with reduced risk among those who were on ART.

An early study where authors compared anal HPV infection in HIV^+ and HIV^- injection drug user women reported that HPV-associated epithelial abnormalities were concomitant with low $CD4^+$ T cell counts (< 200 cells/mL) in peripheral blood and were two times more prevalent as cervical HPV infection in HIV^+ subjects [69,74]. Same results were shown by Zhang et al. in the case of laryngeal squamous cell carcinoma where they reported low ratio of $CD4^+$ and $CD8^+$ cells and found more $CD8^+$ cells infiltrating to tumor site [75]. $CD4^+$ T cells activate $CD8^+$ T cells also referred as cytotoxic T cells (CTLs) to eliminate transformed tumor cells (by lytic and nonlytic pathways) and regulate the immune response [76]. In few cases, spontaneous nongenetic regression of HIV/HPV-associated cervical cancerous cells to precancerous cells has been noticed [77]. A proposed mechanism of development of cancer in HIV/HPV co-infection is shown in Fig. 1. Moreover, reports have indicated that biologically active and extracellular HIV-1 Tat protein is eventually taken up by human uterine

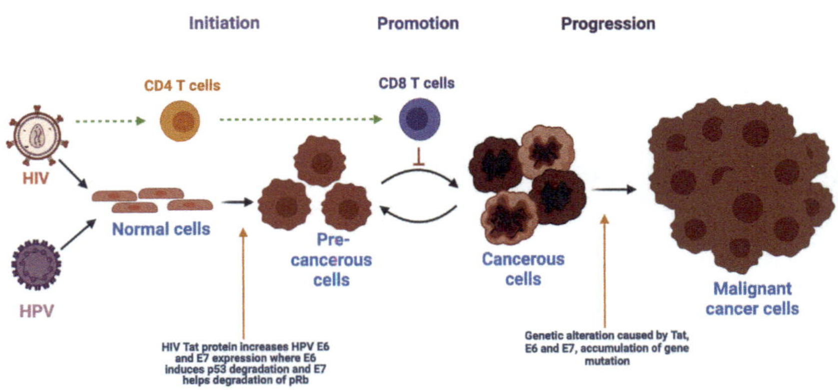

Fig. 1 Pathogenesis of HIV/HPV co-infection-associated CC. The pathogenesis of HR-HPV-induced CC involves rapid proliferation of infected cells and cell cycle deregulation by oncoproteins E6 and E7 to support cell proliferation. The integration of viral genome into host genome results in higher viral protein expression. Ultimately, the accumulation of somatic mutations in host genome leads to transformation. The interaction between HPV/HIV co-infections is also connected with the HIV/HPV-associated CC caused by the persistent buildup and unrepaired mutations. In addition, the HIV and HPV pro-oncogenic proteins Tat, E6, and E7 result in cell transformation which in turn develop into malignant cancerous cells. Compromised functioning of $CD8^+$ T and $CD4^+$ cells also participates in the process. The immune modulation by both HPV and HIV creates an environment for persistent infection to facilitate transformation.

CC ($HR\text{-}HPV^+$) epithelial cells where it alters the p53 protein levels by elevating the expression of HPV E6 and E7 proteins to degrade p53 and phosphorylated retinoblastoma.

4.2 Anogenital warts and vulvar intraepithelial neoplasia

Based on a study, the annual incidences of anogenital warts in the United States are around 160–289 cases per 100,000 persons where 1.2 cases per 1000 women mostly aged between 20 and 24 years [78–80]. Anogenital warts are a common disease, and 90% are caused by nononcogenic HPV types 6 or 11. Warts that occur at certain anatomic sites like under the foreskin and shaft of penis and around the vaginal introitus are called anogenital warts. Warts can also occur at multiple sites on epithelium and tract of cervix, anus, perineum, perianal skin, scrotum, urethra, vagina, etc. The occurrence of intra-anal warts is higher in individuals that have receptive anal intercourse. But, intra-anal warts are also reported in individuals with no history of anal sexual contact.

Like CIN, anogenital warts and vulvar intraepithelial neoplasia (VIN) which is a preinvasive neoplasia of vulva are much more common among HIV^+ females as compared to HIV^- women [81]. Massad et al. studied the prevalence of genital warts and VIN in a cohort of HIV^+ and HIV^- individuals and noticed that the prevalence of warts was more in HIV^+ than in HIV^- subjects. The collective frequency of warts came out to be significantly high in HIV^+ than in seronegative participants. Uncompromised host immunity is vital in the checking of the initial HPV infection that causes genital warts and VIN and in the controlling of cancer development [82].

5. Conclusion

HPV-HIV co-infection has been implicated with high rate of mortality in females due to aggressive CC and thus creates an essential need to clarify the uniqueness of these co-infections, which play pivotal role and serve as the causative factor in the CC and other associated diseases. To develop novel therapeutics against HPV-HIV co-infection, it is crucial to understand the immunological aspects, specifically microenvironment of the infection. Cancer progression is guarded and sustained by altered microenvironment of the lesions and modulation of TLR expression, inefficient presentations of antigens by LCs, and thus ineffective host immunity against the pathogen. HPV-mediated downregulation of NKCs response, ineffective antigen presentation to the CTL, and reduction of Th1 by promoting Th2 and Treg suppress cell-mediated immunity. This knowledge and therapeutic interventions could be helpful in developing and stimulating cell immune response against the pathogens.

References

[1] J. Doorbar, et al., The biology and life-cycle of human papillomaviruses, Vaccine 30 (Suppl 5) (2012) F55–F70.
[2] J.M. Partridge, L.A. Koutsky, Genital human papillomavirus infection in men, Lancet Infect. Dis. 6 (1) (2006) 21–31.
[3] H. Trottier, E.L. Franco, The epidemiology of genital human papillomavirus infection, Vaccine 24 (Suppl 1) (2006) S1–15.
[4] F. Fehrmann, L.A. Laimins, Human papillomaviruses: targeting differentiating epithelial cells for malignant transformation, Oncogene 22 (33) (2003) 5201–5207.
[5] M.S. Campo, et al., HPV-16 E5 down-regulates expression of surface HLA class I and reduces recognition by CD8 T cells, Virology 407 (1) (2010) 137–142.
[6] A. Venuti, et al., Papillomavirus E5: the smallest oncoprotein with many functions, Mol. Cancer 10 (2011) 140.
[7] A.C. de Freitas, et al., hrHPV E5 oncoprotein: immune evasion and related immunotherapies, J. Exp. Clin. Cancer Res. 36 (1) (2017) 71.
[8] U.A. Hasan, et al., The human papillomavirus type 16 E7 oncoprotein induces a transcriptional repressor complex on the Toll-like receptor 9 promoter, J. Exp. Med. 210 (7) (2013) 1369–1387.
[9] E.M. Smith, et al., Evidence for vertical transmission of HPV from mothers to infants, Infect. Dis. Obstet. Gynecol. 2010 (2010) 326369.
[10] V. Mello, R.K. Sundstrom, Cervical intraepithelial neoplasia, in: StatPearls, StatPearls Publishing LLC., Treasure Island, FL, 2021.
[11] U. Sahu, P. Khare, Role of interleukin-17 in human papillomavirus infection and associated malignancies, Microb. Pathog. (2021) 105294.
[12] R. Nayar, D.C. Wilbur, The Pap Test and Bethesda 2014. "The reports of my demise have been greatly exaggerated." (after a quotation from Mark Twain), Acta Cytol. 59 (2) (2015) 121–132.
[13] K.A. Workowski, G.A. Bolan, Sexually transmitted diseases treatment guidelines, 2015, MMWR Recomm. Rep. 64 (RR-03) (2015) 1–137.
[14] Human immunodeficiency virus (HIV), Transfus. Med. Hemother. 43 (3) (2016) 203–222.
[15] A.C. Araújo, et al., Incidence of cervical intraepithelial neoplasia in a cohort of HIV-infected women, Int. J. Gynaecol. Obstet. 117 (3) (2012) 211–216.
[16] U. Irshad, H. Mahdy, T. Tonismae, HIV in Pregnancy, StatPearls [Internet], StatPearls Publishing, Treasure Island, FL, 2022.
[17] R.M. Badial, et al., Detection and genotyping of human papillomavirus (HPV) in HIV-infected women and its relationship with HPV/HIV co-infection, Medicine (Baltimore) 97 (14) (2018), e9545.
[18] S. Joshi, et al., Human papillomavirus infection among human immunodeficiency virus-infected women in Maharashtra, India, Vaccine 32 (9) (2014) 1079–1085.
[19] T.V. Ellerbrock, et al., Incidence of cervical squamous intraepithelial lesions in HIV-infected women, JAMA 283 (8) (2000) 1031–1037.
[20] J.A. Brinkman, et al., Detection of human papillomavirus DNA in urine specimens from human immunodeficiency virus-positive women, J. Clin. Microbiol. 40 (9) (2002) 3155–3161.
[21] A.R. Garbuglia, et al., Frequency and multiplicity of human papillomavirus infection in HIV-1 positive women in Italy, J. Clin. Virol. 54 (2) (2012) 141–146.

[22] N.J. Veldhuijzen, et al., The epidemiology of human papillomavirus infection in HIV-positive and HIV-negative high-risk women in Kigali, Rwanda, BMC Infect. Dis. 11 (2011) 333.
[23] J. Palefsky, Human papillomavirus-related disease in people with HIV, Curr. Opin. HIV AIDS 4 (1) (2009) 52–56.
[24] U. Sahu, et al., Interleukin-17—a multifaceted cytokine in viral infections, J. Cell. Physiol. (2021) 1–2.
[25] A.M. Bedoya, et al., Location and density of immune cells in precursor lesions and cervical cancer, Cancer Microenviron. 6 (1) (2013) 69–77.
[26] D. Song, et al., Effect of human papillomavirus infection on the immune system and its role in the course of cervical cancer, Oncol. Lett. 10 (2) (2015) 600–606.
[27] M.A. Stanley, J.C. Sterling, Host responses to infection with human papillomavirus, Curr. Probl. Dermatol. 45 (2014) 58–74.
[28] L.V. Ronco, et al., Human papillomavirus 16 E6 oncoprotein binds to interferon regulatory factor-3 and inhibits its transcriptional activity, Genes Dev. 12 (13) (1998) 2061–2072.
[29] J.S. Park, et al., Inactivation of interferon regulatory factor-1 tumor suppressor protein by HPV E7 oncoprotein. Implication for the E7-mediated immune evasion mechanism in cervical carcinogenesis, J. Biol. Chem. 275 (10) (2000) 6764–6769.
[30] F. Mota, et al., The antigen-presenting environment in normal and human papillomavirus (HPV)-related premalignant cervical epithelium, Clin. Exp. Immunol. 116 (1) (1999) 33–40.
[31] S.J. Piersma, Immunosuppressive tumor microenvironment in cervical cancer patients, Cancer Microenviron. 4 (3) (2011) 361–375.
[32] D.B. Alves, et al., CD4 and CD8 T lymphocytes and NK cells in the stroma of the uterine cervix of women infected with human papillomavirus, Rev. Soc. Bras. Med. Trop. 43 (4) (2010) 425–429.
[33] L. Feller, et al., HPV modulation of host immune responses, S. Afr. Dent. J. 65 (6) (2010) 266–268.
[34] B. Jiang, M. Xue, Correlation of E6 and E7 levels in high-risk HPV16 type cervical lesions with CCL20 and Langerhans cells, Genet. Mol. Res. 14 (3) (2015) 10473–10481.
[35] U. Hasan, Human papillomavirus (HPV) deregulation of Toll-like receptor 9, Oncoimmunology 3 (1) (2014), e27257.
[36] Y. Zhang, et al., The expression of Toll-like receptor 8 and its relationship with VEGF and Bcl-2 in cervical cancer, Int. J. Med. Sci. 11 (6) (2014) 608–613.
[37] L.S. Hammes, et al., Macrophages, inflammation and risk of cervical intraepithelial neoplasia (CIN) progression—clinicopathological correlation, Gynecol. Oncol. 105 (1) (2007) 157–165.
[38] A.P. Lepique, et al., HPV16 tumor associated macrophages suppress antitumor T cell responses, Clin. Cancer Res. 15 (13) (2009) 4391–4400.
[39] T. Garcia-Iglesias, et al., Low NKp30, NKp46 and NKG2D expression and reduced cytotoxic activity on NK cells in cervical cancer and precursor lesions, BMC Cancer 9 (2009) 186.
[40] L.J.P. Liebenberg, et al., HPV infection and the genital cytokine milieu in women at high risk of HIV acquisition, Nat. Commun. 10 (1) (2019) 5227.
[41] G.T. Spear, et al., Chemokines are present in the genital tract of HIV-seropositive and HIV-seronegative women: correlation with other immune mediators, J. Acquir. Immune Defic. Syndr. Hum. Retrovirol. 18 (5) (1998) 454–459.
[42] J.M. Kriek, et al., Female genital tract inflammation, HIV co-infection and persistent mucosal Human Papillomavirus (HPV) infections, Virology 493 (2016) 247–254.
[43] C. Ao, K. Zeng, The role of regulatory T cells in pathogenesis and therapy of human papillomavirus-related diseases, especially in cancer, Infect. Genet. Evol. 65 (2018) 406–413.

[44] S.H. Averbach, et al., The association between cervical human papillomavirus infection and HIV acquisition among women in Zimbabwe, AIDS 24 (7) (2010) 1035–1042.
[45] L. Osborn, S. Kunkel, G.J. Nabel, Tumor necrosis factor alpha and interleukin 1 stimulate the human immunodeficiency virus enhancer by activation of the nuclear factor kappa B, Proc. Natl. Acad. Sci. U. S. A. 86 (7) (1989) 2336–2340.
[46] K. Behbakht, et al., Role of the vaginal microbiological ecosystem and cytokine profile in the promotion of cervical dysplasia: a case-control study, Infect. Dis. Obstet. Gynecol. 10 (4) (2002) 181–186.
[47] M. Herfs, et al., Mucosal junctions: open doors to HPV and HIV infections? Trends Microbiol. 19 (3) (2011) 114–120.
[48] P. Hubert, et al., Altered α-defensin 5 expression in cervical squamocolumnar junction: implication in the formation of a viral/tumour-permissive microenvironment, J. Pathol. 234 (4) (2014) 464–477.
[49] D. Guerrero, et al., Hypermethylation of the thrombospondin-1 gene is associated with poor prognosis in penile squamous cell carcinoma, BJU Int. 102 (6) (2008) 747–755.
[50] C.F. Houlihan, et al., Human papillomavirus infection and increased risk of HIV acquisition. A systematic review and meta-analysis, AIDS 26 (17) (2012) 2211–2222.
[51] L. de Witte, et al., Langerin is a natural barrier to HIV-1 transmission by Langerhans cells, Nat. Med. 13 (3) (2007) 367–371.
[52] J. Laurson, et al., Epigenetic repression of E-cadherin by human papillomavirus 16 E7 protein, Carcinogenesis 31 (5) (2010) 918–926.
[53] R.G. Ghebre, et al., Cervical cancer control in HIV-infected women: past, present and future, Gynecol. Oncol. Rep. 21 (2017) 101–108.
[54] L. Pantanowitz, P. Michelow, Review of human immunodeficiency virus (HIV) and squamous lesions of the uterine cervix, Diagn. Cytopathol. 39 (1) (2011) 65–72.
[55] H.K. Whitham, et al., A comparison of the natural history of HPV infection and cervical abnormalities among HIV-positive and HIV-negative women in Senegal, Africa, Cancer Epidemiol. Biomarkers Prev. 26 (6) (2017) 886–894.
[56] M. Verma, S. Erwin, V. Abedi, R. Hontecillas, S. Hoops, A. Leber, J. Bassaganya-Riera, S.M. Ciupe, Modeling the mechanisms by which HIV-associated immunosuppression influences HPV persistence at the oral mucosa, PLoS One 12 (1) (2017) e0168133, https://doi.org/10.1371/journal.pone.0168133 e0168133.
[57] G. Barillari, et al., Entrance of the tat protein of HIV-1 into human uterine cervical carcinoma cells causes upregulation of HPV-E6 expression and a decrease in p53 protein levels, Oncol. Lett. 12 (4) (2016) 2389–2394.
[58] M.L. Tornesello, et al., Human immunodeficiency virus type 1 tat gene enhances human papillomavirus early gene expression, Intervirology 36 (2) (1993) 57–64.
[59] B.Y. Alli, et al., HIV modifies the effect of tobacco smoking on oral human papillomavirus infection, J Infect Dis 222 (4) (2020) 646–654.
[60] Human papillomavirus-associated cancers—United States, 2004–2008, MMWR Morb. Mortal. Wkly Rep. 61 (2012) 258–261.
[61] P.V. Chin-Hong, J.M. Palefsky, Natural history and clinical management of anal human papillomavirus disease in men and women infected with human immunodeficiency virus, Clin. Infect. Dis. 35 (9) (2002) 1127–1134.
[62] L.J. Conley, et al., HIV-1 infection and risk of vulvovaginal and perianal condylomata acuminata and intraepithelial neoplasia: a prospective cohort study, Lancet 359 (9301) (2002) 108–113.
[63] A. Ferenczy, et al., Human papillomavirus and HIV coinfection and the risk of neoplasias of the lower genital tract: a review of recent developments, CMAJ 169 (5) (2003) 431–434.
[64] P. Du, Human papillomavirus infection and cervical cancer in HIV+ women, Cancer Treat. Res. 177 (2019) 105–129.

[65] L.M. Wen, et al., Risk factors for the acquisition of genital warts: are condoms protective? Sex. Transm. Infect. 75 (5) (1999) 312–316.
[66] C. Chelimo, et al., Risk factors for and prevention of human papillomaviruses (HPV), genital warts and cervical cancer, J. Inf. Secur. 66 (3) (2013) 207–217.
[67] W. Satmary, et al., Vulvar intraepithelial neoplasia: risk factors for recurrence, Gynecol. Oncol. 148 (1) (2018) 126–131.
[68] J.M. Palefsky, et al., Cervicovaginal human papillomavirus infection in human immunodeficiency virus-1 (HIV)-positive and high-risk HIV-negative women, J. Natl. Cancer Inst. 91 (3) (1999) 226–236.
[69] A.B. Williams, et al., Anal and cervical human papillomavirus infection and risk of anal and cervical epithelial abnormalities in human immunodeficiency virus-infected women, Obstet. Gynecol. 83 (2) (1994) 205–211.
[70] M. Zardast, et al., Assessment of antibacterial effect of garlic in patients infected with *Helicobacter pylori* using urease breath test, Avicenna J. Phytomed. 6 (5) (2016) 495–501.
[71] A.C. de Freitas, et al., Susceptibility to cervical cancer: an overview, Gynecol. Oncol. 126 (2) (2012) 304–311.
[72] A.A. McBride, A. Warburton, The role of integration in oncogenic progression of HPV-associated cancers, PLoS Pathog. 13 (4) (2017), e1006211.
[73] W.I. Al-Daraji, J.H. Smith, Infection and cervical neoplasia: facts and fiction, Int. J. Clin. Exp. Pathol. 2 (1) (2009) 48–64.
[74] N.A. Hessol, et al., Concomitant anal and cervical human papillomavirusV infections and intraepithelial neoplasia in HIV-infected and uninfected women, AIDS 27 (11) (2013) 1743–1751.
[75] Y. Zhang, et al., Enhancing CD8(+) T cell fatty acid catabolism within a metabolically challenging tumor microenvironment increases the efficacy of melanoma immunotherapy, Cancer Cell 32 (3) (2017) 377–391.e9.
[76] N. Gulzar, K.F. Copeland, CD8+ T-cells: function and response to HIV infection, Curr. HIV Res. 2 (1) (2004) 23–37.
[77] N. Coleman, et al., Immunological events in regressing genital warts, Am. J. Clin. Pathol. 102 (6) (1994) 768–774.
[78] T. Hoy, et al., Assessing incidence and economic burden of genital warts with data from a US commercially insured population, Curr. Med. Res. Opin. 25 (10) (2009) 2343–2351.
[79] J.E. Koshiol, S.A. Laurent, J.M. Pimenta, Rate and predictors of new genital warts claims and genital warts-related healthcare utilization among privately insured patients in the United States, Sex. Transm. Dis. 31 (12) (2004) 748–752.
[80] H. Patel, et al., Systematic review of the incidence and prevalence of genital warts, BMC Infect. Dis. 13 (2013) 39.
[81] K.P. Maniar, et al., Coexisting high-grade vulvar intraepithelial neoplasia (VIN) and condyloma acuminatum: independent lesions due to different HPV types occurring in immunocompromised patients, Am. J. Surg. Pathol. 37 (1) (2013) 53–60.
[82] L.S. Massad, et al., Genital warts and vulvar intraepithelial neoplasia, Obstet. Gynecol. 118 (4) (2011) 831–839.

CHAPTER 5

Cervicovaginal microbiota and HPV-induced cervical cancer

Ashish Jain[a], Swadesh Kishore Shrivastava[b], and Linju Joy[a]
[a]Smt. Chandibai Himathmal Mansukhani College, Ulhasnagar-3, Affiliated to University of Mumbai (Maharashtra), Mumbai, Maharashtra, India
[b]Omisure Technologies, Bhopal, Madhya Pradesh, India

1. Introduction

Cervical cancer (CC) is one of the leading gynecological cancers in females. It is the fourth most prevalent cancer and the fourth main reason of mortality among females due to cancer globally. It annually accounts for more than 500,000 cases and 300,000 mortalities worldwide [1,2]. About one-fifth of cancers are linked to diverse infections caused by microorganisms [3]. Nearly all cases of CC (99%) are linked to high-risk human papillomavirus (HR-HPV) infection. Persistent HPV infection in the cervix can cause high-grade squamous intraepithelial lesions (CIN2–3) or CC [4]. Compromised immunity, co-infection of other sexually transmitted infections (STIs), smoking, obesity, family history of CC, a low nutrition diet, use of contraceptive pills, three full-term pregnancies, teenage pregnancy, vaginal dysbiosis or a diverse and altered cervicovaginal microbiota (CVM) are among the other risk factors [5,6].

CVM is a complex microbial population and a key role player in female reproductive health. The composition of CVM is decided by various intrinsic and extrinsic factors. CVM of healthy women is principally dominated by lactobacilli and helps to maintain vaginal eubiosis by maintaining acidic pH and secreting various metabolites and preventing gynecological disorders [7] (Fig. 1).

Disturbance in the optimum composition of CVM and loss in *Lactobacillus* dominancy may result in a more diverse microbial population which is associated with compromised reproductive health. Previous investigations have confirmed that alterations in the CVM may enhance the probability of CC progression [8,9]. Recent investigations have proved the

Immunopathology, Diagnosis, and Treatment of HPV-induced Malignancies
https://doi.org/10.1016/B978-0-323-91797-1.00010-7

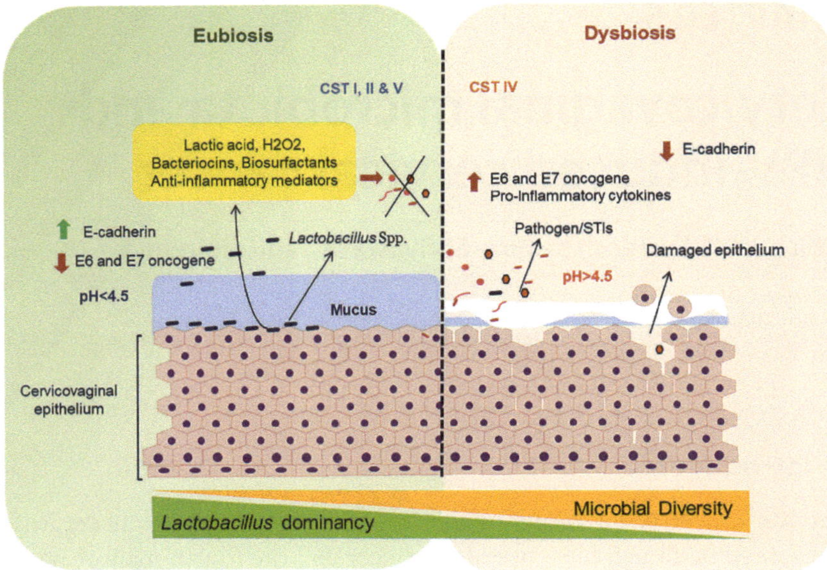

Fig. 1 Cervicovaginal environment during eubiosis and dysbiosis.

role of a healthy CVM in reducing incidences of STIs and HPV-associated cervical malignancies [10]. This book chapter summarized the preventive role of a healthy CVM in HPV infection and HPV-induced cervical malignancy and the consequences of a disturbance in its optimum composition.

2. Human papillomavirus (HPV)

HPVs are called *papillomaviruses* because they can cause papillomas (warts) in the host. HPVs are naked, icosahedral, spherical viruses having a double-stranded DNA of 8000 base pairs and a virion size of around 55 nm. Papillomaviruses are highly epitheliotropic and may cause infection in the mucosal or cutaneous epithelium of different tissues [11]. HPVs replicate in the nucleus of host cells. E6 and E7 are the two main proteins of viral origin that can immortalize and transform host cells by inactivating p53 and retinoblastoma (pRb) proteins respectively associated with tumor suppression [12–15].

2.1 Types of HPV

About 200 different kinds of HPV have been identified based on DNA sequencing [16]. HPV types can be broadly divided into HR-HPV and low-risk HPV (LR-HPV) groups based on their pathogenicity, forms of lesions, and connotation with CC. LR-HPV typically do not cause any serious health issue; however, some can cause benign lesions and warts of genitals, anus, and mouth region. LR-HPVs include type 6, 11, 34, 40, 42, 43, 44, 61, 70, 72, 81, and 89. There are 7 intermediate-risk HPVs including type 26, 53, 66, 67, 84, 90, and 91. There are about 16 HR-HPVs including type 16, 18, 31, 33, 35, 39, 45, 51, 52, 54, 56, 58, 59, 68, 73, and 82 [17]. HR-HPVs can cause higher-grade squamous intraepithelial lesions (HSIL) and several types of malignancies. HPV type 16, 18, 31, 45, 33, 35, 39, 51, 52, 56, 58, 66, 68, and 70 have been frequently reported in cervical carcinoma [18–20]. HPV-16 and HPV18 are accountable for 58.9% and 15% incidences of CC, respectively, the rest are associated with other types of HR-HPVs [21,22].

2.2 HPV and cervical cancer (CC)

HPV is the leading viral infection of the female genital region. HPV infections are more frequent in sexually active women. The majority of cases of HPV were reported soon after becoming sexually active. Although it is an STI, sexual intercourse is not an obligation for spread. Skin-to-skin genital contact is a well-recognized mode of transmission [23]. Warts caused by most types of HPVs are noncancerous tumors. However, certain types of HPVs are known to cause cancers of the cervix, vagina, vulva, penis, anus, and oropharynx [24]. However, in women of low-income countries, it is the most frequently spotted cancer and the main cause of mortality due to cancer. The cervix accounts for 83% of HPV-attributable cancer. More than 4% of incidences of all cancers globally are attributed to HPV infection, females are 10 times more affected than males, 8.6% in females, and 0.8% in males [25]. Although few HPV types are oncogenic, however, HPV is thought to be responsible for almost all CCs. HPV DNA was spotted in more than 90% of CC cases [26]. A persistent HR-HPV infection is crucial for CC development. E6 and E7, are the main proteins of HPV involved in CC development, immortality, and malignancy [27].

3. Cervicovaginal microbiota (CVM)

The human microbiome is a term that describes the genome of all cultivable or noncultivable microorganisms that inhabit and interact with the human body [28]. CVM is considered a commensal, interdependent, and infectious microbial community that inhabits on the cervicovaginal lining. CVM is dominated by bacteria over the fungal (mycobiome) and viral (virome) population.

3.1 Classification of CVM

Vaginal secretions of mature women comprise approximately 10^6–10^8 bacteria per mL of vaginal fluid [29]. In most of the healthy reproductive stage females, this complex CVM is mainly led by specific species of lactobacilli. CVM can be categorized into five diverse community state types (CST) based on the most prevalent and dominant microbial species [30]. Four of these community types CST-I, II, III, and V show dominancy of dissimilar species of *Lactobacillus* namely *L. crispatus*, *L. gasseri*, *L. iners*, *and L. jensenii*, respectively. CST-IV lacks a high proportion of *Lactobacillus* and has a more diverse community of microorganisms mainly composed of anaerobic microorganisms. CST-IV is known for a less acidic environment having pH >4.5, whereas the other four *Lactobacillus*-dominant CSTs exhibit more acidic vaginal pH (<4.5). CST-IV is categorized further into two subtypes: CST-IVA and IVB. CST-IVA shares a modest ratio of lactobacilli, together with a lower number of *Anaerococcus*, *Corynebacterium*, *Finegoldia*, or *Streptococcus*, and CST-IVB has higher proportion of *Atopobium*, along with *Prevotella*, *Parvimonas*, *Sneathia*, *Gardnerella*, *Mobiluncus*, *Peptoniphilus*, and other groups [31]. CST IV-B-related bacteria are a major factor for vaginal disbiosis specified by a reduction of lactobacilli dominancy and abundance of anaerobes. CST-IV microorganisms can disturb the intactness of cervicovaginal epithelium and can initiate a pro-inflammatory immune response along with various other morphological, physiological, and biochemical alterations in cervicovaginal epithelium [32,33].

3.2 CVM and reproductive health

The CVM configuration habitually alters over comparatively short periods. The composition of CVM is influenced by age, diet, smoking, menstruation, hormonal changes, pregnancy, sexual practices, hygiene practices, antimicrobial medication, hormonal contraceptives, and menopause [34–37]. Dominance of *Lactobacillus* in CVM is correlated with cervicovaginal health, and any disturbance in its composition may lead to vaginal dysbiosis and increased incidence of bacterial vaginosis (BV), vulvovaginal candidiasis (VVC), trichomoniasis, HPV, *Chlamydia trachomatis* (CT), human immunodeficiency virus (HIV) vulnerability [38–41]. These infections are linked with different gynecological problems, adverse effects on reproduction, increased occurrences of risk of acquiring STIs, pelvic inflammatory disease, endometritis, premature birth, spontaneous abortions, and cervical malignancies [42]. Many recent investigations have proved the crucial role of cervicovaginal microbiota in maintaining good reproductive health.

Diversity in vaginal microbiota results higher frequency of HR-HPV infection and development of CC [43–45]. HPV infection harmed the normal CVM and increased the microbial diversity of the reproductive tract which further led to cervical malignancies [46].

3.3 Preventive role of healthy CVM in CC

Lactobacillus is a key player in maintaining vaginal eubiosis. It utilizes glycogen of vaginal epithelium and adheres to the cervicovaginal mucosa and competes for available space with other pathogens. They reduce cervicovaginal inflammation, secrete lactic acid, H_2O_2, other acidic metabolites, antimicrobial bacteriocins, biosurfactants and make a hostile environment for dysbiotic microbes and establish vaginal eubiosis [47–49]. They prevent biofilm formation by urogenital pathogens and their adhesion to cervicovaginal mucosa [50]. Lactobacilli maintain noninflammatory conditions in the reproductive tract and suppress the proinflammatory cytokine facilitating the employment of cells targeted by HIV and reducing the risk of STIs while the presence of diverse CVM is associated with a proinflammatory environment that enhances the risk of STIs [51]. Lactobacilli can assist in repair of vaginal tissue, reduce HPV infection and also contribute to the clearance of HPV infection [1,52]. *Lactobacillus* can accelerate apoptosis of CC cell line HeLa hence can be included in preventive and therapeutic strategies for CC [53]. Many metabolic intermediates secreted by *Lactobacillus* can reduce the progression of tumors [54]. Culture supernatant of lactobacilli reduced proliferation of Caski cell, reduced the level of HPV oncogenes E6 and E7, regulators of cell cycle CDK2 and cyclin-A as well as arrested most of the cells in S-phase compared to G2/M phase [55]. *L. plantarum* isolated from vaginal fluid of healthy women exert anticancer potential against human CC cell line HeLa without any cytotoxic effect toward normal host cells [56]. Exopolysaccharides produced by vaginal isolate *L. gasseri*, inhibited proliferation of HeLa cells and induced its apoptosis by upregulating Bax and Caspase 3 [57]. Supernatant of lactobacilli can prevent metastasis and proliferation of human CC cells HeLa by downregulating the expression of Caspase 3, matrix metalloproteinases-2 (MMP2), and MMP9 [53]. *Bifidobacterium adolescentis* SPM1005-A can suppress E6 and E7 oncogene expression of HR-HPV-16 hence can be a potential candidate to prevent HPV-associated CC [58]. Lactic acid produced by lactobacilli can reduce the expression of E6 and E7 oncogenes and inhibit cell movement and invasion of Sinha CC cells through upregulating E-cadherin in tumor tissues [59,60]. E-cadherin has a crucial role in contact inhibition and its suppression results in loss of contact inhibition, increased cell motility, and advanced phases of malignancy [61]. Vaginal lactic acid bacteria exert cytotoxic effects on the cancerous Hela cell line, with considerable safety toward normal fibroblast-like cervical cells [62]. Loss in lactobacillus dominancy can lead to increased CVM diversity and growth of the mixed population of anaerobic bacteria. Harmful metabolites produced by these diverse bacterial species in combination with other factors including HPV infection can create the optimum environment for various gynecological syndromes including cervical malignancies [38,63].

4. CVM, sexually transmitted infections (STIs), and CC

It has been proved that a persistent HPV infection is among the main causes of CC [64]. However other decisive factors of CC are not well understood. Recent investigations indicate an important role of CVM in the acquisition and establishment of HPV infection and consequent progression of cervical malignancies. An altered ratio of the CVM and the occurrence of other infections are linked with an augmented risk of CC. Latest investigations have revealed complicated interactions among the CVM and STIs that probably decide the progression of CC [65,66]. In the majority of women, a stable and healthy vaginal environment show comparatively less diverse flora mainly dominated by *Lactobacillus* species [30,31,43]. A drop in *Lactobacillus* dominancy arouses diversity in CVM and ampleness of anaerobic bacterial species and a higher risk for STIs and cervicovaginal disorders [67].

4.1 HPV infection and CC

Metagenomic studies revealed a highly diverse and non-*Lactobacillus* dominant CVM in HPV$^+$ women. A more diverse and non-*Lactobacillus* dominant CVM was found in women with CIN and CC than normal women [30,43,68–70]. In a 5-year study, more than 70% of women diagnosed with CC showed disturbances in the normal composition of CVM [71]. Nondominancy of lactobacilli is related to 3–5 times greater risk of HPV infection and 2–3 times higher threat for HR-HPV and CC compared with dominancy of *L. crispatus* [72]. The decreased population of *L. crispatus* and a higher ratio of dysbiotic bacteria including *Atopobium vaginae*, *Dialister invisus*, *F. magna*, *G. vaginalis*, *Prevotella buccalis*, and *Prevotella timonensis* have been reported in CIN 1, 2 or 3, and CC [43]. HPV is linked with a reduced level of probiotic bacteria and an increased population of pathogens [73]. Patients who recovered with HR-HPV infection restored *L. crispatus* dominancy and reduced level of inflammatory cytokines compared to HR-HPV persistent patients [74].

4.2 Bacterial vaginosis (BV) and CC

BV is a vaginal syndrome among reproductive-age women where dominant lactobacilli growth is suppressed and they are replaced by a more diverse bacterial community, mainly anaerobes consisting of *Actinomyces, Anaerococcus Atopobium vagina*, *Bacteroides* spp., *Dialister, G. vaginalis, Megasphaera, Mobiluncus* spp., *Mycoplasma hominis, Peptoniphilus, Peptostreptococcus,* and *Prevotella* spp., *Slackia, Streptococcus, Ureaplasma* [75–78]. BV is characterized by a fishy odor of cervicovaginal secretions, pH more than 4.5, a confirmative amine "whiff" test, and existence of clue cells in vaginal discharge broadly known as "*Amsel criteria*". Untreated cases of BV may lead to gynecological complications, including genital inflammation, genitourinary tract infections, and increased risk of cervical HPV, HIV, and other nonviral STDs [78–80]. Enzymes secreted by the diverse bacterial population of BV degrade cervicovaginal mucin and disrupt protective mucosal barrier, which results in vulnerability to cervicovaginal HPV infection, attachment, penetration of HPV, and integration of its genetic material. Anaerobic bacteria of BV may also secrete carcinogenic ammonia nitrite, causing abnormality in cervical epithelial cells [81].

More diversity in CVM is reported in patients with precancerous CIN and invasive cervical carcinoma (ICC) [38,82]. Several investigations proved that the prevalence of BV-related bacteria in the course of HR-HPV co-infection can increase the risk for CC [75]. Infections of *Atopobium vaginae, D. invisus, F. magna, G. vaginalis, Prevotella buccalis, Ureaplasma,* and *Prevotella timonensis* were linked with the occurrence of high-grade squamous intraepithelial neoplasia and HPV associated CC [43,83,84]. *Atopobium vaginae* is a proven risk factor of CIN3 and its dominancy in CVM is an important risk factor for the development of cervical neoplasia [85–87]. The abundance of *Prevotella* is related to a persistent HPV infection and has an inverse correlation with the dominancy of *Lactobacillus* [88]. *Sneathia* is linked with HPV infection, cervical precancer, and cancer [38,82,89].

4.3 *Trichomonas vaginalis* and CC

Trichomoniasis, triggered by the eukaryotic, extracellular, flagellated, facultatively anaerobic protozoan *T. vaginalis*, is one of the predominant nonviral STI globally. Several studies have revealed that women with a preexisting abnormal CVM or with a Nugent score of more than three showed a higher risk of getting TV infection [90–92]. *Lactobacillus* spp. have been reported to reduce the growth rate of TV in vitro [93]. Nondominancy of *Lactobacillus* spp. has been reported in women having persistent TV infection [94]. *T. vaginalis* along with CST-IV type bacteria cooperatively decreases the integrity of the tight junctions of cervicovaginal epithelium and makes it more permeable for other STIs [95]. Aggregation-promoting factor APF-2 of lactobacilli inhibits cytoadherence of *T. vaginalis* to cervicovaginal epithelium cells [96]. Dysbiotic CVM acts as pathobionts for *T. vaginalis* infection as they enhance the cytoadherence of trophozoites and enhance their pathogenicity [97].

Women infected with *T. vaginalis* have a considerably greater risk of developing cervical abrasions and CC specifically when co-infected with HPV than those without exposure [98,99]. The role of *T. vaginalis* in cervical malignancy is not fully elucidated however it is reported to disrupt of cervicovaginal mucosa and allow HPV to penetrate the basal layer of the epithelium. As a result, it leads to the integration of viral DNA into the host DNA and the overexpression of viral oncogenes that contribute to the activation of carcinogenic mechanisms [100,101].

4.4 *Chlamydia trachomatis* and CC

A healthy vaginal microbiota dominated by *Lactobacilli* adversely affects adsorption and replication of *Chlamydia trachomatis* [102]. Metabolites of *Lactobacilli* can reduce the infectivity of *Chlamydia* infection by acting on its elementary body [103]. *Lactobacillus* can reduce infectivity of *Chlamydia trachomatis* by modify cervicovaginal epithelial cell membranes through altering lipid composition and α5 integrin modulation [104]. *Lactobacillus* also reduces inflammation caused by the pathogen [105]. Exceptionally *L. iners* dominant CVM is linked with greater risk for *Chlamydia trachomatis* infection [106].

Chlamydia trachomatis is a gram-negative, obligate intracellular bacterial STI that infects cells of the reproductive tract. Women having Chlamydial infection are at greater risk of acquiring a persistent HPV infection that may further lead to cervical intraepithelial neoplasia [107,108]. Chlamydial infection results in a disrupted epithelium and an altered immune response which facilitate HPV entry and obstruct virus clearance [109,110]. Increased cell proliferation was observed in cervical tissue of mice infected with *Chlamydia trachomatis* [111]. Women living with *Chlamydia trachomatis* infection have more chances of cervical squamous cell carcinomas. *Chlamydia* prevents apoptosis in the host cell, alters normal pathways, misfolds host proteins, induces mutation rate and alters DNA repair mechanism in host cells during infection these strategies may induce malignancies in infected tissue [112,113].

4.5 Human immunodeficiency virus (HIV) and CC

Lactobacillus-rich CVM is associated with reduced incidences of HIV infection. Lactic acid secreted by lactobacilli can inactivate HIV and induce host cervicovaginal epithelium to produce antiinflammatory cytokines and inhibit recruitment of HIV target cells [114,115]. Lactobacilli secret bacteriocins, H_2O_2, and other anti-HIV metabolites and maintain thick cervicovaginal mucus which acts as a barrier and can trap HIV [51]. Loss of *Lactobacillus* dominancy and diverse anaerobic bacterial communities in the female reproductive tract (FRT) can increase HIV acquisition by up to 60% by recruiting $CD4^+$ mucosal immune cells the main target for HIV [116–118]. Abnormal CVM is associated with the decreased population of protective, cervical gamma delta 1 (GD1) cells, and an increased number of GD2 cells as targets for HIV entry [119]. Mannose-specific lectins of lactobacilli can bind with HIV gp120 and block the membrane fusion process required for HIV-1 entry [120,121]. HIV-inducing factor (HIF) which induces HIV replication by triggering activator protein 1 (AP-1) and nuclear factor kappa B (NF-κB) in host cells were frequently detected in the vagina of women infected with dysbiotic microbiota, especially with *Mycoplasma hominis* [122].

Although HPV and HIV are different viruses, both are considered STIs and can remain dormant for years without causing any symptoms in the host. HPV is known to cause genital warts; a co-infection of HIV can increase the disease severity and there is a possibility of abnormal cervicovaginal cells which can develop into certain cancers [123,124]. Around 77% HIV^+ women also carry HPV [125]. Tat and gp120 proteins of HIV are reported to separate tight junctions and increase the chances of HPV interacting with basal epithelial cells which are the target cells for HPV-induced malignancies [126]. HPV can also increase the risk of HIV infection. E7 protein of HPV can reduce E-cadherin expression in infected tissues which leads to a higher susceptibility of cervicovaginal lining for HIV infection [127]. T-lymphocytes produced in response to HPV infection serve as target cells for HIV and favor to establish HIV infection [128]. Females having an HIV infection have six times more chances to develop CC. Worldwide, about 6% of fresh CC incidences in 2018 were diagnosed in HIV-positive females and 5% of them were caused due to HIV infection [129]. A higher risk to acquire HR-HPV infection and the lower possibility to clear infection is a common phenomenon in women co-infected with HIV [130]. Immune suppression and low titer of $CD4^+$ cells due to HIV indirectly increase the severity of HPV infection, the possibility of oncogenesis, and the reappearance of postmedication HPV infection [131]. A higher mortality rate was recorded in those CC patients who had an HIV co-infection in addition to HPV [132].

5. Restoration of CVM to prevent CC

A diverged CVM can be modulated by oral or local administrations of a *Lactobacillus* spp. dominant probiotic formulation. Oral administration of probiotic Lactobacilli effectively restores healthy cervicovaginal flora and positively influences female reproductive health [133,134]. Probiotics have been found effective in the prevention and controlling of BV and other STIs [135]. The application of local vaginal probiotics can decrease the risk of HPV infection and increase the rate of its clearance. This can reduce the incidences of genital malignancies and morbidity [46,136]. Restored CVM excludes attachment of pathogenic bacteria and competes with them for nutrition and colonization of the vagina and may play a vital role in immunomodulation of tumor microenvironment in the host [133]. Further investigation is required to conclude the impact of modulation of CVM using probiotic supplements as a preventive or curative approach in the treatment of gynecologic malignancies.

6. Conclusion

CC is triggered by an amalgamation of hereditary elements and external environmental stimuli. Although persistent infection with oncogenic HR-HPV is a major element that leads to CC, but it is not enough. The composition of CVM could be a deciding factor in the progression or prevention of HPV infection and gynecological malignancies. Disturbance in CVM composition could be a risk factor whereas an optimally balanced microbiota may reduce the risk of genital disorders including CC. The composition of CVM can decide the direction of the molecular pathway followed by HR-HPV infected targeted cells. CVM positively influences female reproductive health by several known and unknown mechanisms. The decisive role of CVM in cervical malignancies can be predicted by analysis of the exact composition of the microbiota of healthy individuals and cancer patients.

References

[1] M. Arbyn, et al., Worldwide burden of cervical cancer in 2008, Ann. Oncol. 22 (12) (Dec. 2011) 2675–2686.

[2] F. Bray, J. Ferlay, I. Soerjomataram, R.L. Siegel, L.A. Torre, A. Jemal, Global cancer statistics 2018: GLOBOCAN estimates of incidence and mortality worldwide for 36 cancers in 185 countries, CA Cancer J. Clin. 68 (6) (2018) 394–424.

[3] D. Elsland, J. Neefjes, Bacterial infections and cancer, EMBO Rep. 19 (11) (2018).

[4] R. Kawahara, et al., Changes to the cervicovaginal microbiota and cervical cytokine profile following surgery for cervical intraepithelial neoplasia, Sci. Rep. 11 (1) (2021) 2156.

[5] R. Herrero, R. Murillo, Cervical cancer, in: M.J. Thun, M.S. Linet, J.R. Cerhan, C.A. Haiman (Eds.), Cancer Epidemiology and Prevention, fourth ed., Oxford University Press, New York, 2018.

[6] X. Castellsagué, Natural history and epidemiology of HPV infection and cervical cancer, Gynecol. Oncol. 110 (3) (2008) S4–S7.

[7] S. Boris, C. Barbés, Role played by lactobacilli in controlling the population of vaginal pathogens, Microbes Infect. 2 (5) (2000) 543–546.

[8] M. Kwon, S.S. Seo, M.K. Kim, D.O. Lee, M.C. Lim, Compositional and functional differences between microbiota and cervical carcinogenesis as identified by shotgun metagenomic sequencing, Cancers (Basel) 11 (3) (2019).

[9] S. Arokiyaraj, S.S. Seo, M. Kwon, J.K. Lee, M.K. Kim, Association of cervical microbial community with persistence, clearance and negativity of Human Papillomavirus in Korean women: a longitudinal study, Sci. Rep. 8 (1) (2018) 15479.

[10] V. Verhoeven, et al., Probiotics enhance the clearance of human papillomavirus-related cervical lesions, Eur. J. Cancer Prev. 22 (1) (2013) 46–51.

[11] IARC Working Group on the Evaluation of Carcinogenic Risks to Humans. Human Papillomaviruses. Lyon (FR): International Agency for Research on Cancer, "(IARC monographs on the evaluation of carcinogenic risks to humans, No. 90.) 1, Human Papillomaviru," 2007, 2007, [Online]. Available https://www.ncbi.nlm.nih.gov/books/NBK321770/. (Accessed 29 October 2021).

[12] E. Langsfeld, L.A. Laimins, Human papillomaviruses: research priorities for the next decade, Trends Cancer 2 (5) (May 2016) 234–240.

[13] O. Basukala, L. Banks, The not-so-good, the bad and the ugly: HPV E5, E6 and E7 oncoproteins in the orchestration of carcinogenesis, Viruses 13 (10) (2021) 1892.

[14] S.V. Graham, The human papillomavirus replication cycle, and its links to cancer progression: a comprehensive review, Clin. Sci. 131 (17) (2017) 2201–2221.

[15] K. Münger, P.M. Howley, Human papillomavirus immortalization and transformation functions, Virus Res. 89 (2) (2002) 213–228.

[16] H.Z. Hausen, Papillomaviruses in human cancers, Proc. Assoc. Am. Physicians 111 (6) (1999) 581–587.

[17] M.P. Bihl, et al., Human papillomavirus (HPV) detection in Cytologic specimens: similarities and differences of available methodology, Appl. Immunohistochem. Mol. Morphol. 25 (3) (2017) 184–189.

[18] V. Bouvard, et al., A review of human carcinogens—part B: biological agents, Lancet Oncol. 10 (4) (2009) 321–322.

[19] E.M. Burd, Human papillomavirus and cervical cancer, Clin. Microbiol. Rev. 16 (1) (2003) 1–17.

[20] F.X. Bosch, et al., Prevalence of human papillomavirus in cervical cancer: a worldwide perspective, JNCI J. Natl. Cancer Inst. 87 (11) (1995) 796–802.

[21] N. Muñoz, et al., Epidemiologic classification of human papillomavirus types associated with cervical Cancer, N. Engl. J. Med. 348 (6) (2003) 518–527.

[22] M. Stanley, Pathology and epidemiology of HPV infection in females, Gynecol. Oncol. 117 (2) (2010) S5–S10.
[23] H. Sung, et al., Global Cancer statistics 2020: GLOBOCAN estimates of incidence and mortality worldwide for 36 cancers in 185 countries, CA Cancer J. Clin. 71 (3) (2021) 209–249.
[24] D. Forman, et al., Global burden of human papillomavirus and related diseases, Vaccine 30 (2012) F12–F23.
[25] C. de Martel, M. Plummer, J. Vignat, S. Franceschi, Worldwide burden of cancer attributable to HPV by site, country and HPV type, Int. J. Cancer 141 (4) (2017) 664–670.
[26] M. Saraiya, et al., US assessment of HPV types in cancers: implications for current and 9-valent HPV vaccines, JNCI J. Natl. Cancer Inst. 107 (6) (2015).
[27] S.M. Wilting, R.D.M. Steenbergen, Molecular events leading to HPV-induced high grade neoplasia, Papillomavirus Res. 2 (Dec. 2016) 85–88.
[28] P.J. Turnbaugh, R.E. Ley, M. Hamady, C.M. Fraser-Liggett, R. Knight, J.I. Gordon, The human microbiome project, Nature 449 (7164) (2007) 804–810.
[29] D. Danielsson, P.K. Teigen, H. Moi, The genital econiche: focus on microbiota and bacterial vaginosis, Ann. N.Y. Acad. Sci. 1230 (1) (Aug. 2011) 48–58.
[30] J. Ravel, et al., Vaginal microbiome of reproductive-age women, Proc. Natl. Acad. Sci. U. S. A. 108 (Suppl._1) (Mar. 2011) 4680–4687.
[31] P. Gajer, et al., Temporal dynamics of the human vaginal microbiota, Sci. Transl. Med. 4 (132) (2012) 132ra52.
[32] F. Bradley, et al., The vaginal microbiome amplifies sex hormone-associated cyclic changes in cervicovaginal inflammation and epithelial barrier disruption, Am. J. Reprod. Immunol. 80 (1) (2018), e12863.
[33] L.-J. Sierra, et al., Colonization of the cervicovaginal space with *Gardnerella vaginalis* leads to local inflammation and cervical remodeling in pregnant mice, PLoS One 13 (1) (2018), e0191524.
[34] R.J. Hickey, X. Zhou, J.D. Pierson, J. Ravel, L.J. Forney, Understanding vaginal microbiome complexity from an ecological perspective, Transl. Res. 160 (4) (2012) 267–282.
[35] R. Romero, et al., The composition and stability of the vaginal microbiota of normal pregnant women is different from that of non-pregnant women, Microbiome 2 (1) (2014) 4.
[36] K.L. Nunn, L.J. Forney, Unraveling the dynamics of the human vaginal microbiome, Yale J. Biol. Med. 89 (3) (2016) 331–337.
[37] A.L. Muhleisen, M.M. Herbst-Kralovetz, Menopause and the vaginal microbiome, Maturitas 91 (Sep. 2016) 42–50.
[38] A. Mitra, et al., Cervical intraepithelial neoplasia disease progression is associated with increased vaginal microbiome diversity, Sci. Rep. 5 (1) (2015) 16865.
[39] J.H.H.M. van de Wijgert, The vaginal microbiome and sexually transmitted infections are interlinked: Consequences for treatment and prevention, PLoS Med. 14 (12) (2017), e1002478.
[40] N. Ziklo, M.E. Vidgen, K. Taing, W.M. Huston, P. Timms, Dysbiosis of the vaginal microbiota and higher vaginal kynurenine/tryptophan ratio reveals an association with *Chlamydia trachomatis* genital infections, Front. Cell. Infect. Microbiol. 8 (2018).
[41] M.C. Eastment, R.S. McClelland, Vaginal microbiota and susceptibility to HIV, AIDS 32 (6) (2018) 687–698.
[42] D.H. Martin, J.M. Marrazzo, The vaginal microbiome: current understanding and future directions, J Infect Dis 214 (Suppl 1) (2016) S36–S41.
[43] K.A. So, et al., Changes of vaginal microbiota during cervical carcinogenesis in women with human papillomavirus infection, PLoS One 15 (9) (2020), e0238705.
[44] Y. Chen, et al., Human papillomavirus infection and cervical intraepithelial neoplasia progression are associated with increased vaginal microbiome diversity in a Chinese cohort, BMC Infect. Dis. 20 (1) (2020) 629.

[45] X.-P. Chao, et al., Correlation between the diversity of vaginal microbiota and the risk of high-risk human papillomavirus infection, Int. J. Gynecol. Cancer 29 (1) (2019) 28–34.
[46] M. Kyrgiou, A. Mitra, A.-B. Moscicki, Does the vaginal microbiota play a role in the development of cervical cancer? Transl. Res. 179 (2017) 168–182.
[47] D. Nasioudis, I. Linhares, W. Ledger, S. Witkin, Bacterial vaginosis: a critical analysis of current knowledge, BJOG Int. J. Obstet. Gynaecol. 124 (1) (2017) 61–69.
[48] X.-X. Niu, T. Li, X. Zhang, S.-X. Wang, Z.-H. Liu, Lactobacillus crispatus modulates vaginal epithelial cell innate response to Candida albicans, Chin Med J (Engl) 130 (3) (2017) 273–279.
[49] A.A. Medina-Colorado, et al., Vaginal ecosystem modeling of growth patterns of anaerobic bacteria in microaerophilic conditions, Anaerobe 45 (2017) 10–18.
[50] M.I. Petrova, et al., The lectin-like protein 1 in *Lactobacillus rhamnosus* GR-1 mediates tissue-specific adherence to vaginal epithelium and inhibits urogenital pathogens, Sci. Rep. 6 (1) (2016) 37437.
[51] H. Borgdorff, et al., Cervicovaginal microbiome dysbiosis is associated with proteome changes related to alterations of the cervicovaginal mucosal barrier, Mucosal Immunol. 9 (3) (2016) 621–633.
[52] K. Takada, et al., *Lactobacillus crispatus* accelerates re-epithelialization in vaginal epithelial cell line MS74, Am. J. Reprod. Immunol. 80 (3) (2018), e13027.
[53] Z. Nouri, et al., Dual anti-metastatic and anti-proliferative activity assessment of two probiotics on HeLa and HT-29 cell lines, Cell J. 18 (2) (2016) 127–134.
[54] P. Zadravec, B. Štrukelj, A. Berlec, Improvement of LysM-mediated surface display of designed Ankyrin repeat proteins (DARPins) in recombinant and nonrecombinant strains of Lactococcus lactis and Lactobacillus species, Appl. Environ. Microbiol. 81 (6) (2015) 2098–2106.
[55] K.-D. Wang, D.-J. Xu, B.-Y. Wang, D.-H. Yan, Z. Lv, J.-R. Su, Inhibitory effect of vaginal Lactobacillus supernatants on cervical Cancer cells, Probiotics Antimicrob. Proteins 10 (2) (2018) 236–242.
[56] Y. Nami, N. Abdullah, B. Haghshenas, D. Radiah, R. Rosli, A.Y. Khosroushahi, Assessment of probiotic potential and anticancer activity of newly isolated vaginal bacterium Lactobacillus plantarum 5BL, Microbiol. Immunol. 58 (9) (2014) 492–502.
[57] T. Sungur, B. Aslim, C. Karaaslan, B. Aktas, Impact of exopolysaccharides (EPSs) of Lactobacillus gasseri strains isolated from human vagina on cervical tumor cells (HeLa), Anaerobe 47 (2017) 137–144.
[58] M.-K. Cha, et al., Antiviral activity of *Bifidobacterium adolescentis*SPM1005-A on human papillomavirus type 16, BMC Med. 10 (1) (2012) 72.
[59] C. Li, L. Jia, Y. Yu, L. Jin, Lactic acid induced microRNA-744 enhances motility of SiHa cervical cancer cells through targeting ARHGAP5, Chem. Biol. Interact. 298 (2019) 86–95.
[60] X. Li, et al., Lactobacilli inhibit cervical cancer cell migration in vitro and reduce tumor burden in vivo through upregulation of E-cadherin, Oncol. Rep. 38 (3) (2017) 1561–1568.
[61] A.M. Mendonsa, T.-Y. Na, B.M. Gumbiner, E-cadherin in contact inhibition and cancer, Oncogene 37 (35) (2018) 4769–4780.
[62] E. Motevaseli, et al., The effect of Lactobacillus crispatus and Lactobacillus rhamnosusCulture supernatants on expression of autophagy genes and HPV E6 and E7 oncogenes in the HeLa cell line, Cell J. 17 (4) (2016) 601–607.
[63] E.O. DARENG, et al., Prevalent high-risk HPV infection and vaginal microbiota in Nigerian women, Epidemiol. Infect. 144 (1) (2016) 123–137.
[64] H. De Vuyst, et al., The burden of human papillomavirus infections and related diseases in sub-Saharan Africa, Vaccine 31 (2013) F32–F46.

[65] F. Godoy-Vitorino, G. Ortiz-Morales, J. Romaguera, M.M. Sanchez, M. Martinez-Ferrer, N. Chorna, Discriminating high-risk cervical Human Papilloma Virus infections with urinary biomarkers via non-targeted GC-MS-based metabolomics, PLoS One 13 (12) (2018), e0209936.
[66] C. Klein, C. Kahesa, J. Mwaiselage, J.T. West, C. Wood, P.C. Angeletti, How the cervical microbiota contributes to cervical cancer risk in Sub-Saharan Africa, Front. Cell. Infect. Microbiol. 10 (2020).
[67] S.J. Kroon, J. Ravel, W.M. Huston, Cervicovaginal microbiota, women's health, and reproductive outcomes, Fertil. Steril. 110 (3) (2018) 327–336.
[68] W. Kwasniewski, et al., Microbiota dysbiosis is associated with HPV-induced cervical carcinogenesis, Oncol. Lett. 16 (6) (2018) 7035–7047.
[69] F. De Seta, G. Campisciano, N. Zanotta, G. Ricci, M. Comar, The vaginal community state types microbiome-immune network as key factor for bacterial vaginosis and aerobic vaginitis, Front. Microbiol. 10 (2019).
[70] R. Wakabayashi, Y. Nakahama, V. Nguyen, J.L. Espinoza, The host-microbe interplay in human papillomavirus-induced carcinogenesis, Microorganisms 7 (7) (2019) 199.
[71] S.M. Kovachev, Cervical cancer and vaginal microbiota changes, Arch. Microbiol. 202 (2) (2020) 323–327.
[72] J. Norenhag, J. Du, M. Olovsson, H. Verstraelen, L. Engstrand, N. Brusselaers, The vaginal microbiota, human papillomavirus and cervical dysplasia: a systematic review and network meta-analysis, BJOG Int. J. Obstet. Gynaecol. 127 (2) (2020) 171–180.
[73] J. Liu, M. Luo, Y. Zhang, G. Cao, S. Wang, Association of high-risk human papillomavirus infection duration and cervical lesions with vaginal microbiota composition, Ann. Transl. Med. 8 (18) (2020) 1161.
[74] E. Caselli, et al., Vaginal microbiota and cytokine microenvironment in HPV clearance/persistence in women surgically treated for cervical intraepithelial neoplasia: An observational prospective study, Front. Cell. Infect. Microbiol. 10 (2020).
[75] Z.-T. Wei, H.-L. Chen, C.-F. Wang, G.-L. Yang, S.-M. Han, S.-L. Zhang, Depiction of vaginal microbiota in women with high-risk human papillomavirus infection, Front. Public Health 8 (2021).
[76] L. Monin, E.M. Whettlock, V. Male, Immune responses in the human female reproductive tract, Immunology 160 (2) (2020) 106–115.
[77] A. Georgijević, S. Cjukić-Ivancević, M. Bujko, Bacterial vaginosis. Epidemiology and risk factors, Srp. Arh. Celok. Lek. 128 (1–2) (2000) 29–33.
[78] E. Gillet, et al., Bacterial vaginosis is associated with uterine cervical human papillomavirus infection: a meta-analysis, BMC Infect. Dis. 11 (1) (2011) 10.
[79] H. Verstraelen, R. Verhelst, Bacterial vaginosis: an update on diagnosis and treatment, Expert Rev. Anti Infect. Ther. 7 (9) (2009) 1109–1124.
[80] Biswal, Current concept of bacterial vaginosis in cervical cancer, J. Clin. Gynecol. Obstet 3 (2014) 1–7.
[81] Y. Liang, M. Chen, L. Qin, B. Wan, H. Wang, A meta-analysis of the relationship between vaginal microecology, human papillomavirus infection and cervical intraepithelial neoplasia, Infect. Agent. Cancer 14 (1) (2019) 29.
[82] A. Audirac-Chalifour, et al., Cervical microbiome and cytokine profile at various stages of cervical cancer: A pilot study, PLoS One 11 (4) (2016), e0153274.
[83] A.M. Ekiel, D.A. Friedek, M.K. Romanik, J. Jóźwiak, G. Martirosian, Occurrence of *Ureaplasma parvum* and *Ureaplasma urealyticum* in women with cervical Dysplasia in Katowice, Poland, J. Korean Med. Sci. 24 (6) (2009) 1177.
[84] H.Y. Oh, et al., The association of uterine cervical microbiota with an increased risk for cervical intraepithelial neoplasia in Korea, Clin. Microbiol. Infect. 21 (7) (2015) 674.e1–674.e9.
[85] N. Brusselaers, S. Shrestha, J. van de Wijgert, H. Verstraelen, Vaginal dysbiosis and the risk of human papillomavirus and cervical cancer: systematic review and meta-analysis, Am. J. Obstet. Gynecol. 221 (1) (2019) 9–18.e8.

[86] C.S. Bradshaw, S.N. Tabrizi, C.K. Fairley, A.N. Morton, E. Rudland, S.M. Garland, The Association of Atopobium vaginae and Gardnerella vaginalis with bacterial vaginosis and recurrence after Oral metronidazole therapy, J Infect Dis 194 (6) (2006) 828–836.
[87] S.-S. Seo, H.Y. Oh, J.-K. Lee, J.-S. Kong, D.O. Lee, M.K. Kim, Combined effect of diet and cervical microbiome on the risk of cervical intraepithelial neoplasia, Clin. Nutr. 35 (6) (2016) 1434–1441.
[88] J. Si, H.J. You, J. Yu, J. Sung, G. Ko, Prevotella as a hub for vaginal microbiota under the influence of host genetics and their association with obesity, Cell Host Microbe 21 (1) (2017) 97–105.
[89] M. Di Paola, et al., Characterization of cervico-vaginal microbiota in women developing persistent high-risk Human Papillomavirus infection, Sci. Rep. 7 (1) (2017) 10200.
[90] J.E. Balkus, et al., Bacterial vaginosis and the risk of trichomonas vaginalis acquisition among HIV-1–negative women, Sex. Transm. Dis. 41 (2) (2014) 123–128.
[91] S.D. Rathod, K. Krupp, J.D. Klausner, A. Arun, A.L. Reingold, P. Madhivanan, Bacterial vaginosis and risk for trichomonas vaginalis infection: a longitudinal analysis, Sex. Transm. Dis. 38 (9) (2011) 882–886.
[92] R.M. Brotman, et al., Bacterial vaginosis assessed by gram stain and diminished colonization resistance to incident gonococcal, chlamydial, and Trichomonal genital infection, J Infect Dis 202 (12) (2010) 1907–1915.
[93] J.G. Rendón-Maldonado, M. Espinosa-Cantellano, A. Gonzalez-Robles, A. Martınez-Palomo, Trichomonas vaginalis: in VitroPhagocytosis of lactobacilli, vaginal epithelial cells, leukocytes, and erythrocytes, Exp. Parasitol. 89 (2) (1998) 241–250.
[94] M.R. Torok, et al., The association between trichomonas vaginalis infection and level of vaginal lactobacilli, in nonpregnant women, J Infect Dis 196 (7) (2007) 1102–1107.
[95] A.S. Hinderfeld, N. Phukan, A.-K. Bär, A.M. Roberton, A. Simoes-Barbosa, Cooperative interactions between trichomonas vaginalis and associated bacteria enhance paracellular permeability of the cervicovaginal epithelium by dysregulating tight junctions, Infect. Immun. 87 (5) (2019).
[96] N. Phukan, A.E.S. Brooks, A. Simoes-Barbosa, A cell surface aggregation-promoting factor from *Lactobacillus gasseri* contributes to inhibition of Trichomonas vaginalis adhesion to human vaginal ectocervical cells, Infect. Immun. 86 (8) (2018).
[97] A.S. Hinderfeld, A. Simoes-Barbosa, Vaginal dysbiotic bacteria act as pathobionts of the protozoal pathogen trichomonas vaginalis, Microb. Pathog. 138 (2020) 103820.
[98] R.-Y. Su, et al., Association between trichomonas vaginalis infection and cervical lesions: a population-based, nested case-control study in Taiwan, Parasitol. Res. 119 (8) (2020) 2649–2657.
[99] S. Yang, W. Zhao, H. Wang, Y. Wang, J. Li, X. Wu, Trichomonas vaginalis infection-associated risk of cervical cancer: a meta-analysis, Eur. J. Obstet. Gynecol. Reprod. Biol. 228 (2018) 166–173.
[100] F. Mercer, P.J. Johnson, Trichomonas vaginalis: pathogenesis, symbiont interactions, and host cell immune responses, Trends Parasitol. 34 (8) (2018) 683–693.
[101] I. Nikas, et al., Integrated morphologic and molecular analysis of trichomonas vaginalis, mycoplasma hominis, and human papillomavirus using cytologic smear preparations, Parasitol. Res. 117 (5) (2018) 1443–1451.
[102] P. Mastromarino, M. Di Pietro, G. Schiavoni, C. Nardis, M. Gentile, R. Sessa, Effects of vaginal lactobacilli in Chlamydia trachomatis infection, Int. J. Med. Microbiol. 304 (5–6) (2014) 654–661.
[103] P. Nardini, et al., *Lactobacillus crispatus* inhibits the infectivity of *Chlamydia trachomatis* elementary bodies, in vitro study, Sci. Rep. 6 (1) (2016) 29024.
[104] C. Parolin, et al., *Lactobacillus crispatus* BC5 interferes with *Chlamydia trachomatis* infectivity through integrin modulation in cervical cells, Front. Microbiol. 9 (2018).
[105] A. Rizzo, M. Fiorentino, E. Buommino, G. Donnarumma, A. Losacco, N. Bevilacqua, Lactobacillus crispatus mediates anti-inflammatory cytokine interleukin-10 induction

in response to Chlamydia trachomatis infection in vitro, Int. J. Med. Microbiol. 305 (8) (2015) 815–827.
[106] R. van Houdt, B. Ma, S.M. Bruisten, A.G.C.L. Speksnijder, J. Ravel, H.J.C. de Vries, Lactobacillus iners -dominated vaginal microbiota is associated with increased susceptibility to Chlamydia trachomatis infection in Dutch women: a case–control study, Sex. Transm. Infect. 94 (2) (2018) 117–123.
[107] K.E. Jensen, et al., *Chlamydia trachomatis* and risk of cervical intraepithelial neoplasia grade 3 or worse in women with persistent human papillomavirus infection: a cohort study, Sex. Transm. Infect. 90 (7) (2014) 550–555.
[108] M. Safaeian, et al., Chlamydia trachomatis and risk of prevalent and incident cervical Premalignancy in a population-based cohort, JNCI J. Natl. Cancer Inst. 102 (23) (2010) 1794–1804.
[109] C.A. Horvath, G.A. Boulet, V.M. Renoux, P.O. Delvenne, J.-P.J. Bogers, Mechanisms of cell entry by human papillomaviruses: an overview, Virol. J. 7 (1) (2010) 11.
[110] A.C. Simonetti, J. de Humberto Lima Melo, P.R.E. de Souza, D. Bruneska, J. de Luiz Lima Filho, Immunological's host profile for HPV and Chlamydia trachomatis, a cervical cancer cofactor, Microbes Infect. 11 (4) (2009) 435–442.
[111] A.E. Knowlton, L.J. Fowler, R.K. Patel, S.M. Wallet, S.S. Grieshaber, Chlamydia induces anchorage independence in 3T3 cells and detrimental cytological defects in an infection model, PLoS One 8 (1) (2013), e54022.
[112] M.A. Al-Zeer, et al., Chlamydia trachomatis prevents apoptosis via activation of PDPK1-MYC and enhanced mitochondrial binding of hexokinase II, EBioMedicine 23 (2017) 100–110.
[113] C. Chumduri, R.K. Gurumurthy, P.K. Zadora, Y. Mi, T.F. Meyer, Chlamydia infection promotes host DNA damage and proliferation but impairs the DNA damage response, Cell Host Microbe 13 (6) (2013) 746–758.
[114] M. Aldunate, et al., Vaginal concentrations of lactic acid potently inactivate HIV, J. Antimicrob. Chemother. 68 (9) (2013) 2015–2025.
[115] A.C. Hearps, et al., Vaginal lactic acid elicits an anti-inflammatory response from human cervicovaginal epithelial cells and inhibits production of pro-inflammatory mediators associated with HIV acquisition, Mucosal Immunol. 10 (6) (2017) 1480–1490.
[116] J. Atashili, C. Poole, P.M. Ndumbe, A.A. Adimora, J.S. Smith, Bacterial vaginosis and HIV acquisition: a meta-analysis of published studies, AIDS 22 (12) (2008) 1493–1501.
[117] C. Gosmann, et al., Lactobacillus -deficient Cervicovaginal bacterial communities are associated with increased HIV Acquisition in Young South African Women, Immunity 46 (1) (2017) 29–37.
[118] H. Borgdorff, et al., Lactobacillus-dominated cervicovaginal microbiota associated with reduced HIV/STI prevalence and genital HIV viral load in African women, ISME J. 8 (9) (2014) 1781–1793.
[119] M.L. Alcaide, et al., Bacterial vaginosis is associated with loss of gamma delta T cells in the female reproductive tract in women in the miami women interagency HIV Study (WIHS): A cross sectional study, PLoS One 11 (4) (2016), e0153045.
[120] S. Malik, et al., High mannose-specific lectin Msl mediates key interactions of the vaginal *Lactobacillus plantarum* isolate CMPG5300, Sci. Rep. 6 (1) (2016) 37339.
[121] M.R. Boyd, et al., Discovery of cyanovirin-N, a novel human immunodeficiency virus-inactivating protein that binds viral surface envelope glycoprotein gp120: potential applications to microbicide development, Antimicrob. Agents Chemother. 41 (7) (1997) 1521–1530.
[122] J.A. Cohn, et al., HIV-inducing factor in Cervicovaginal secretions is associated with bacterial vaginosis in HIV-1-infected women, JAIDS J. Acquir. Immune Defic. Syndr. 39 (3) (2005) 340–346.

[123] J. Musa, et al., HIV and development of epithelial cell abnormalities in women with prior normal cervical cytology in Nigeria, Infect. Agent. Cancer 15 (1) (2020) 50.
[124] A. Madan, S. Patil, L. Nakate, A study of pap smear in HIV-positive females, J. Obstet. Gynecol. India 66 (6) (2016) 453–459.
[125] K.O. Myers, N.U. Ahmed, The role of HIV in the progression through the stages of the human papillomavirus to cervical cancer pathway, AIDS Rev. 20 (2) (2019).
[126] D. Konopnicki, S. De Wit, N. Clumeck, HPV and HIV coinfection: a complex interaction resulting in epidemiological, clinical and therapeutic implications, Future Virol. 8 (9) (2013) 903–915.
[127] J. Laurson, S. Khan, R. Chung, K. Cross, K. Raj, Epigenetic repression of E-cadherin by human papillomavirus 16 E7 protein, Carcinogenesis 31 (5) (2010) 918–926.
[128] L.J.P. Liebenberg, et al., HPV infection and the genital cytokine milieu in women at high risk of HIV acquisition, Nat. Commun. 10 (1) (2019) 5227.
[129] D. Stelzle, et al., Estimates of the global burden of cervical cancer associated with HIV, Lancet Glob. Health 9 (2) (2021) e161–e169.
[130] G. Liu, M. Sharma, N. Tan, R.V. Barnabas, HIV-positive women have higher risk of human papilloma virus infection, precancerous lesions, and cervical cancer, AIDS 32 (6) (2018) 795–808.
[131] P. Debeaudrap, J. Sobngwi, P.-M. Tebeu, G.M. Clifford, Residual or recurrent precancerous lesions after treatment of cervical lesions in human immunodeficiency virus–infected women: a systematic review and Meta-analysis of treatment failure, Clin. Infect. Dis. 69 (9) (2019) 1555–1565.
[132] S. Dryden-Peterson, et al., HIV infection and survival among women with cervical Cancer, J. Clin. Oncol. 34 (31) (2016) 3749–3757.
[133] R.R. Spurbeck, C.G. Arvidson, Lactobacilli at the front line of defense against vaginally acquired infections, Future Microbiol. 6 (5) (2011) 567–582.
[134] L. Petricevic, F.M. Unger, H. Viernstein, H. Kiss, Randomized, double-blind, placebo-controlled study of oral lactobacilli to improve the vaginal flora of postmenopausal women, Eur. J. Obstet. Gynecol. Reprod. Biol. 141 (1) (2008) 54–57.
[135] G. Vujic, A. Jajac Knez, V. Despot Stefanovic, V. Kuzmic Vrbanovic, Efficacy of orally applied probiotic capsules for bacterial vaginosis and other vaginal infections: a double-blind, randomized, placebo-controlled study, Eur. J. Obstet. Gynecol. Reprod. Biol. 168 (1) (2013) 75–79.
[136] M. Kyrgiou, G. Koliopoulos, P. Martin-Hirsch, M. Arbyn, W. Prendiville, E. Paraskevaidis, Obstetric outcomes after conservative treatment for intraepithelial or early invasive cervical lesions: systematic review and meta-analysis, Lancet 367 (9509) (2006) 489–498.

CHAPTER 6

Current diagnostic tools for HPV

Aneesh Sethu Madhavan[a], Deepti Joshi[b], Tanya Sharma[b], Shashwati Nema[a], and Debasis Biswas[a]
[a]Department of Microbiology, All India Institute of Medical Sciences Bhopal, Bhopal, Madhya Pradesh, India
[b]Department of Pathology & Laboratory Medicine, All India Institute of Medical Sciences Bhopal, Bhopal, Madhya Pradesh, India

1. Introduction

Human papillomavirus (HPV) is a highly abundant sexually transmitted pathogen across the world and is reported to cause >90% cases of anal and cervical cancers (CC), about 70% cases of vaginal and vulvar cancers, and >60% cases of penile cancers. Apart from its role in anogenital neoplasms, recent studies demonstrated that about 70% of oropharynx cancers are also linked with HPV [1]. The load of HPV infection shows a significant geographic variation with the prevalence of infection and the related disease being significantly higher in developing countries. While the age-standardized incidence rates (ASIR) for HPV-related cancer were stated to be 6·9 cases per 100,000 person-years in high-income countries, while 16·1 in low-income countries [2].

The natural course of HPV infection and disease progression is insidious and chronic. The virus infects the stratified squamous epithelial cells by gaining access to the basal keratinocytes through a microtrauma of the surface mucosa. The viral replication, depending on the host replication machinery, is synchronized with the differentiation of the epithelial layers and culminates in the passive release of infectious viral particles during the disintegration and desquamation of the superficial epithelial layers. Though the majority of the infections remain transient and are cleared by the host immunity in course of time, persistent infection causes cellular transformation and leads to invasive cancer, which is orchestrated by the continued expression of viral oncogenes, especially E6 and E7 [3].

HPVs are grouped into high-risk (HR) and low-risk (LR) types based on the oncogenic potential. Of more than many genotypes currently recognized among the HPVs, 14 are considered HR types and include the

Immunopathology, Diagnosis, and Treatment of HPV-induced Malignancies
https://doi.org/10.1016/B978-0-323-91797-1.00009-0

types 16, 18, 31, 33, 35, 39, 45, 51, 52, 56, 58, 59, 66, and 68. Among the HR types, 16 and 18 are associated with most of the HPV-related cancers. The remaining genotypes are low risk and are either nonpathogenic or associated with benign anogenital and cutaneous warts [4,5]. There is a wide preponderance of HPV infections among all sexually active individuals and the life-threatening consequences in some of them. This underscores the need for detecting precancerous lesions and early-stage cancers that can provide opportunities for early administration of appropriate therapeutic interventions. The diagnostic modalities for HPV infection have broadly focused on cytology, histology, and molecular techniques and screening programs incorporating these diagnostic techniques have been a major component of population-level cancer prevention strategies.

This chapter aims to present an update on the various diagnostic techniques and screening programs for HPV infection worldwide in both resource-rich and resource-poor countries.

2. Diagnostic techniques for HPV infection

2.1 Sampling strategies

HPV is an intracellular pathogen, which has no viremic phase and remains restricted to differentiated epithelial cells. Hence, the choice of sample for HPV detection focuses on the collection of cellular materials for cytology, biopsy, and/or nucleic acid detection [6]. In the case of cervical lesions, the usual samples include cervical scrapings or brushings. For noncervical neoplasms, in which the role of HPV is being increasingly recognized, invasive tissue samples are usually collected for diagnostic workup [7]. However, technological advances have resulted in the introduction of liquid-based collection devices, that offer certain distinct advantages [7,8]. To facilitate sample collection further, the World Health Organization (WHO) has recommended the use of self-collection devices as part of self-care interventions to improve CC screening, and several countries have evaluated the performance of self-collection devices in improving screening processes [8–11].

Cervical scrapings/brushings, using brush, broom, spatula, tampon, lavage, and swab, are usually collected for cytology and/or nucleic acid testing. The transformation zone, which represents the transition zone between columnar and squamous epithelium, is the site for initiation of dysplastic changes associated with cervical malignancy, and its location changes with age. Sampling strategies are aimed to cover this region along with the ectocervix and endocervix [12].

In the conventional method of collection of cervical scrapings/brushings, the patient is placed on an examination table either in a dorsal lithotomy position or Sims' position. After observing the cervix by inserting a speculum, the cells from the ectocervix are obtained by rotating a wooden or plastic ectocervical spatula. A separate endocervical brush or broom is inserted into the cervix and rotated to ensure the removal of an adequate number of cells from the endocervix. The spatula is smeared onto a slide, immediately followed by rolling the endocervical brush on the same slide and fixed with fixatives like 95% ethyl alcohol. Given the complexity of the procedure, the conventional method of sample collection is done in healthcare settings by trained professionals [13].

Currently, liquid-based cytology (LBC) collection devices are being increasingly used, which come packaged as self-contained kits containing the collection tools (spatula and broom/brush) and a liquid medium for suspending the cellular material collected. After the sample is collected in a conventional manner, the material is dispersed in an alcoholic medium containing either ethanol or methanol [8,14]. The LBC devices offer the advantages of better cytology results by enabling the preparation of a thin smear with a monolayer of cells and allowing the use of the same sample for nucleic acid testing. Characteristics of FDA-approved LBC devices are depicted in Table 1 [7].

An abnormal result in cytology screening is an indication for referral for colposcopic examination and eventual biopsy of abnormal regions of the cervix. Using a colposcope improves the illumination and provides magnification for better visualization of the cervix, especially the transformation zone. Abnormal areas are identified by sequentially applying ordinary saline, 3%–5% Lugol's iodine, and acetic acid. After the patient is positioned on the examination table in a dorsal lithotomy position, a speculum is inserted to facilitate visualization of the vagina and cervix. Application of normal saline by spraying or with cotton balls primarily aids in assessing the blood vessels of the cervix and borders of the transformation zone. Next, the application of 3%–5% acetic acid further helps in assessing the squamocolumnar junction and detecting abnormal areas of the transformation zone through acetowhite changes. Application of Lugol's iodine stains the tissue in shades of brown depending on the amount of glycogen present. Abnormal transformation zones contain very little glycogen and hence appear pale. Finally, the Lugol's iodine also delineates the borders of abnormal tissue better, thus facilitating biopsy [15]. A prediction of the histologic diagnosis may be attempted by using a modified Reid colposcopic index (RCI). The parameters like color, lesion margin and surface configuration, vessels, and iodine-staining findings are scored numerically and interpreted as follows (Table 2) [16].

Table 1 Self-contained LBC devices approved by FDA.

Name of the LBC device	Manufacturer	Medium	Tests performed	Volume	Stability	Year of FDA approval
ThinPrep	Hologic Inc.	Methanol-based PreservCyt	Cytology and nucleic acid testing from the same vial (sampling device discarded)	20 mL vial	Room temperature up to 14 days; 4°C up to 3 months; freezing not recommended	1996
SurePath	Becton Dickinson	Ethanol-based SurePath	Separate vial for cytology and nucleic acid testing (sampling device retained)	5 mL vial		1999

Table 2 Modified reid colposcopic index (RCI) [16].

Colposcopic signs	Zero point	One point	Two points
Color	Acetowhitening: assessed as transparent or translucent, overshoots the margin of the transformation zone, pure white with intense surface shine	Gray or white with a shiny surface	Dull opaque, gray
Lesion margin and surface configuration	Contour: microcondylomatous or micropapillary Flat lesions with indistinct margins Satellite lesions beyond the transformation zone	Symmetrical regular-shaped lesions with straight and smooth outlines	Edges rolled or peeling Internal demarcations among areas of different colposcopic appearance—a peripheral region of low-grade change and central region of high-grade change
Vessels	Fine/uniform-caliber vessels: narrowly and evenly placed Poorly formed configurations of fine punctation and/or mosaic Vessels outside the border of the transformation zone Fine vessels within microcondylomatous or micropapillary lesions	Absent vessels	Sharply demarcated—and widely and randomly placed
Iodine staining	Mahogany brown color Negative uptake of irrelevant lesion, i.e., yellow staining via a lesion scoring three points or less on the first three criteria Areas outside the border of the transformation zone, visible on colposcopy, evident as iodine-negative areas	Partial iodine uptake—variegated, speckled Appearance	Negative iodine uptake of a significant lesion, i.e., yellow staining by a lesion already scoring four points or more on the first three criteria
Overall RCI score		Likely histologic classification	
0–2		CIN 1	
3–5		Overlapping lesion: CIN 1 or CIN 2	
5–8		CIN 2–3	

After identifying the abnormal regions, multiple punch biopsies are taken from the abnormal areas closest to the zone of transformation and sent for a histopathology examination. An endocervical curettage may also be performed simultaneously. Alternatively, the cervical biopsies may be snap-frozen in an isopentane bath or placed in normal saline and frozen. These biopsy samples can also be used for nucleic acid testing. Tissue samples from other anogenital and oropharyngeal lesions can also be similarly collected for histopathology examination and nucleic acid testing [17].

Self-sampling kits for HPV testing, consisting of brush, swab, lavage, or tampon, are available, and WHO has recommended using them to improve the implementation of CC screening in places with challenging healthcare infrastructure and poor availability of trained healthcare providers [18]. Many of the kits contain a swab or brush that may be used by the individual to collect samples from the vagina. The sample collected is transferred as per the user-friendly kit instructions and transported to a testing lab usually at room temperature.

A recent metaanalysis has shown that samples collected by self-collection devices and processed by PCR-based nucleic acid detection methods had comparable sensitivities to samples collected by healthcare workers for lesions with grades of CIN 2 and CIN 3. The same metaanalysis has further found that self-collected samples had lower sensitivity compared to the latter in non-PCR-based methods and point-of-care methods like *care*-HPV [19].

2.2 Cytology

Cervical cytology also called as Papanicolaou test or Pap smear has been used as a modality of screening women for detecting precancerous lesions or early-stage malignancies of the cervix since the 1950s. Cytology has also been used to detect HPV infections of the penis, vulva, and anus. The sample collection for cytology testing has already been described earlier. In conventional cytology, the slides that are smeared with the cervical sample collected with both the spatula and brush/broom are stained and observed under a light microscope for cytological changes. The presence of modified squamous epithelial cells known as koilocytes on microscopic examination of cytology smears is considered the hallmark of HPV infection. The cytopathic effects of HPV include the development of halos around the nucleus with a densely stained cytoplasm peripheral rim, nuclear atypia with irregular nuclear outline along with hyperchromasia, and multinucleation. These changes are collectively called koilocytosis.

Papanicolaou and Babes introduced the use of exfoliative cytology to identify women with CC in the 1920s. Later in the 1950s, conventional cytology was used to identify precancerous lesions also. Over a period of time, several terminologies were used for describing the cytology findings that include Papanicolaou classes (classes 1–5), WHO terminologies (mild, moderate, severe dysplasia, carcinoma in situ, microinvasive carcinoma, and invasive carcinoma), and CIN terminologies (CIN 1, 2, 3, invasive carcinoma). With a better understanding of CC pathogenesis and to avoid confusion among various terminologies, the Bethesda system, which was first introduced in the United States in 1988 and last updated in 2014, has been widely adopted for reporting the cytology results. The Bethesda system reports include comments on the appropriateness of the sample, general categorization, descriptive diagnosis, and management guidelines [20,21].

The abnormalities of epithelial cells that are reported as per 2014 Bethesda system for cervical lesions include both squamous and glandular. The squamous cell abnormalities are classified into 4 categories: (i) atypical squamous cells, (ii) low-grade squamous intraepithelial lesions (LSIL), (iii) high-grade squamous intraepithelial lesions (HSIL), and (iv) squamous cell carcinoma (SCC). ASC refers to cytological changes that suggest SIL but are quantitatively or qualitatively insufficient for a definite diagnosis. These atypical squamous cells are further described as either "atypical squamous cells—undetermined significance" (ASC-US) or "atypical squamous cells—cannot exclude an HSIL (ASC-H)." ASC-US refers to changes that resemble LSIL characterized by an enlarged nucleus, a small increase in nuclear-to-cytoplasmic ratio, minimal hyperchromasia, atypical parakeratosis (parakeratotic cells with denser eosinophilic cytoplasm and enlarged regular nuclei), or incomplete koilocytosis. For ASC-US, cytology reflex HPV testing is recommended. ASC-H refers to lesions that resemble HSIL but the cells are sparse. The cells are smaller, similar to the size of a metaplastic cell, with a high nuclear-to-cytoplasmic ratio along with a focally irregular nucleus, hyperchromasia, and irregular nuclear chromatin. Colposcopy is suggested for women with ASC-H cytology. LSIL implies that the predominant cells are of the superficial or intermediate type with enlarged nuclei that may be hyperchromatic, but the nuclear chromatin is regular, nucleolus is inconspicuous, and cytoplasm is mature and well defined. The cells may be multinucleated and show a variable contour of the nuclear membrane.

The presence of koilocytosis along with a nuclear abnormality falls in the LSIL category. The presence of cells with dense eosinophilic cytoplasm and increased keratinization also indicates an LSIL lesion. HSIL is typically characterized by an increased nucleus-to-cytoplasm ratio. The cells are generally smaller with less cytoplasmic maturity as compared to LSIL. Nucleoli are usually absent but the nuclear membrane is irregular. The cytoplasm may be lacy or dense metaplastic. The dysplastic cells are generally seen in syncytial clusters or small groups (Fig. 1).

It is not uncommon to have smears with morphological features of both HSIL (moderate-to-severe dysplasia) and LSIL (mild dysplasia).

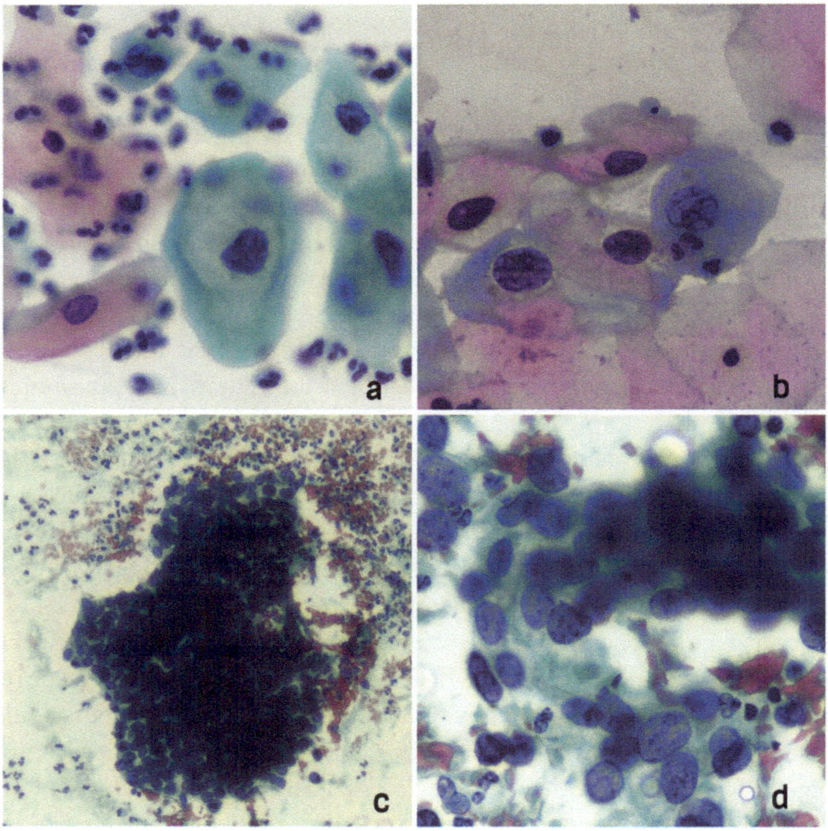

Fig. 1 Cervicovaginal Pap smear showing koilocytic change and intermediate-type squamous cells with enlarged hyperchromatic nuclei and mature well-defined cytoplasm in LSIL ((A, B) 40 × Papanicolaou); hyperchromatic crowded group in HSIL ((C) 10 × Papanicolaou); smaller cells with hyperchromatic nuclei having a high NC ratio, irregular nuclear membrane, and less amount of cytoplasm in HSIL ((D) 40 × Papanicolaou).

SCC of the cervix may be of the keratinizing and nonkeratinizing types. Keratinizing carcinomas show cellular pleomorphism and the presence of caudate and spindle cells which have dense orangeophilic cytoplasm. The cells are generally dispersed and have prominent nucleoli. The non-keratinizing carcinoma has abundant malignant cells arranged loose and forming syncytia. There is a characteristic background containing necrotic material and blood (tumor diathesis) that differentiates carcinoma from HSIL. This background is more pronounced in nonkeratinizing carcinoma [15] (Fig. 2).

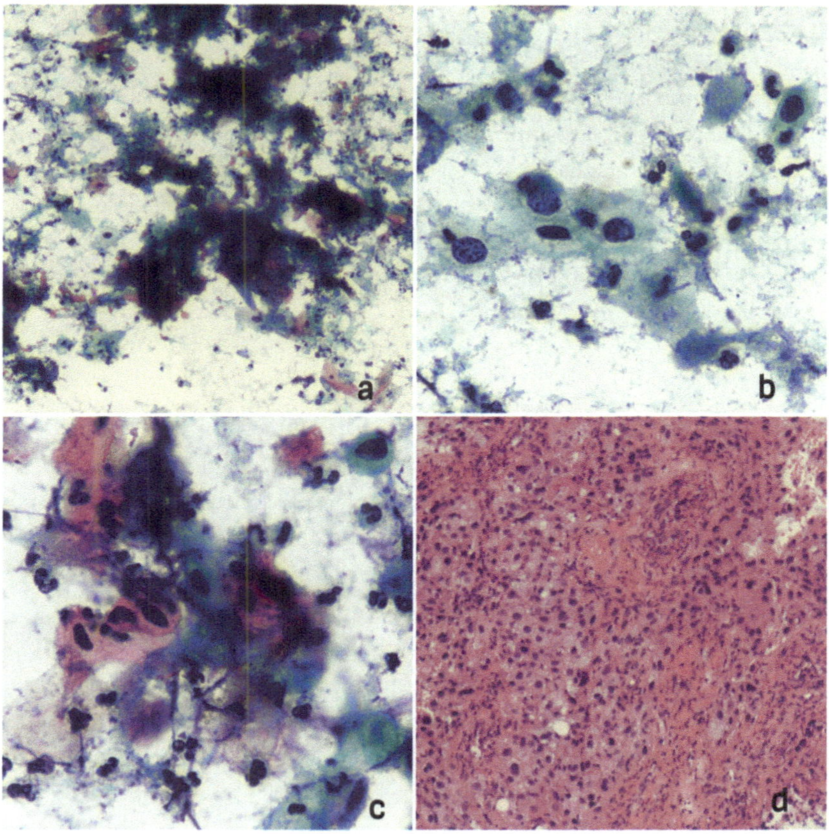

Fig. 2 Cervicovaginal Pap smear in a case of squamous cell carcinoma showing atypical squamous epithelial cells in a background of tumor diathesis ((A) 10× Papanicolaou); Atypical caudate and spindle-shaped squamous epithelial cells showing dense orangeophilic cytoplasm ((B, C) 40× Papanicolaou); Histopathological section of cervical biopsy showing moderately differentiated squamous cell carcinoma ((D) 40× H&E).

Along with reporting of squamous cell abnormalities, the Bethesda system of cytologic reporting also includes a study of glandular cell abnormalities described as per the following scheme:

(a) Atypical endocervical cells (NOS).
(b) Atypical endometrial cells (NOS).
(c) Atypical glandular cells (NOS).
(d) Atypical endocervical cells (favor neoplastic).
(e) Atypical glandular cells (favor neoplastic).
(f) Endocervical adenocarcinoma in situ.
(g) Endocervical adenocarcinoma.
(h) Endometrial adenocarcinoma.
(i) Extrauterine adenocarcinoma.
(j) Adenocarcinoma, not otherwise specified (NOS).

Atypical glandular cells with endocervical origin are seen in aggregates with overlapping nuclei. In cases interpreted as *atypical glandular cells—favor neoplasia,* there is a greater number of abnormal cells. In cases of adenocarcinoma in situ, more cells with crowded clusters arranged in sheets are seen. Varying sizes of endocervical cells with atypical nuclei and coarse clumped chromatin arranged in rosettes are identifying features. Adenocarcinoma cells have high nuclear-to-cytoplasmic ratios, coarsely clumped chromatin, and single or clustered nucleoli [22].

2.3 Histology

There are several indicators for a benign HPV infection of the skin: acanthosis, parakeratosis, and hyperkeratosis involving various layers of the skin like stratum granulosum, stratum spinosum, and stratum corneum respectively being the predominant ones. Although histologic criteria are available for HPV-associated diseases of other anatomic sites, the squamous cell changes in the cervix have been studied in detail.

The sample collection for histopathology testing using a colposcope has already been described earlier. The biopsy sample is subjected to observation under a light microscope following staining for various histopathological changes. Intraepithelial neoplasia is appreciated as the abnormal basal cell proliferation along with the existence of koilocytes. The extent of basal cell-type proliferation determines the grade of the lesion. While proliferation up to the lower third of the epithelial layer denotes mild dysplasia or CIN 1, proliferation up to the upper two-thirds indicates moderate dysplasia or CIN 2. Severe dysplasia/carcinoma in situ (CIS) or CIN 3 is termed when proliferation is observed in the entire breadth of the epithelium. Finally, in invasive SCC, the local break of the basal membrane is seen. Other features that characterize abnormal cellular proliferation and thus denote the severity of lesion include coarse nucleus granularity, elevated cell nucleus/cytoplasm ratio, irregular mitotic figures, and necrosed cells. Interestingly, in higher grades, anisocytosis is seen and koilocytosis may be characteristically absent [16].

The following table demonstrates the mutual correlates between the cytologic and histopathologic grades, which has got clinical implications in deciding the choice of treatment. The terms LSIL and HSIL are interchangeably being used in reporting both cytology and histopathology (Table 3) [23].

Difficulties in accurately determining the histological finding of precancerous lesions, especially CIN 2, have directed the development of several cytological and histological markers associated with progression to malignancy [24]. Protocols involving the application of heat and chemical treatment are available for exposing the antigens, following which the commercially available antibodies could be used to detect the same. Cellular or viral proteins denoting higher oncogenic risk can be detected. Among them, the most extensively studied marker is p16INK4A, which is a cellular protein marker of the family of cyclin-dependent kinase (CDK) inhibitors that are increasingly expressed due to HPV E7 protein-mediated deprivation of the retinoblastoma gene, leading to increased transcription of p16 gene. p16INK4A immune staining has been used on both cytological smear as well as biopsy samples for improved characterization of CIN2 lesions [25]. According to lower anogenital squamous terminology (LAST) project recommendations of 2012, a p16INK4A^{+} lesion would be known as HSIL, while a p16INK4A^{-} lesion is considered LSIL [26]. The advantages of biomarkers include quicker detection of abnormal cells and better reproducibility. Other cellular proteins that have been targeted to detect progression to malignancy include Ki-67 which is a marker for cell proliferation [27]. The viral markers that indicate the onset of cancer progression include the E4, E6, E7, and L1 viral proteins.

Table 3 Correlation between cytological and histological findings [23].

Cytology finding	Histology finding
LSIL	Mild dysplasia/CIN 1
HSIL	Moderate and severe dysplasia, CIS; CIN 2 and CIN 3

In addition, visual inspection using 3%–5% acetic acid and Lugol's iodine have been evaluated for CC screening, especially in resource-limited settings [28].

2.4 Nucleic acid-based detection

Molecular methods for the detection of HPV have led to the dramatic evolution in the discovery of new types, identifying associations with new diseases, epidemiological studies, easing diagnosis in patient care, and vaccine program monitoring. The primary goal of these assays is to detect HPV DNA for screening of cervical samples due to the high sensitivity and negative predictive value [29]. Other goals of molecular tests include reflex testing of specimens diagnosed as ASC-US to decide on the need for colposcopy and co-testing with cytology in women aged 30 years or older [30]. These have been incorporated into the screening programs of various countries. In addition to various in-house HPV tests, around 193 molecular detection assays belonging to 8 groups have been commercialized predominantly for a qualitative assay [31]. Most assays are based on nucleic acid amplification technology (NAAT) and require validation prior to approval for clinical use. Most assays are target amplification based, mainly involving polymerase chain reaction (PCR) of L1 ORF and E6 ORF of viral DNA genome. Other techniques include transcription-mediated amplification (TMA) and nucleic acid sequence-based amplification (NASBA) for RNA targets. mRNA methods are assumed to be more specific for clinically significant infections as these show the presence of actively replicating viruses. The utility of genotyping and availability of automated platforms have gone a long way in revolutionizing HPV detection in lesions at various anatomic sites [23].

Nucleic acid detection tests are either indicated for use as "primary HPV screening tests" or as part of "co-testing" in screening programs, in which these tests are performed along with cytology. Currently, two commercial kits are FDA approved as primary HPV tests, while seven are approved for co-testing. The key aspects of various FDA-approved nucleic acid detection methods for cervical lesions are discussed here.

The COBAS HPV test (Roche Molecular Systems, Inc., Pleasanton, CA) is a real-time PCR assay centered on TaqMan chemistry, which targets the L1 region of the HPV genome and detects high-risk types 16, 18, 31, 33, 35, 39, 45, 51, 52, 56, 58, 59, 66, and 68. This assay is available on an automated platform as COBAS 4800 HPV. The amplicon incorporates dUTP instead of dTTP allowing the deprivation of any *contaminant* amplicons with uracil-*N*-glycosidase (which will not destruct the thymidine-containing target DNA). It uses the globin gene to ascertain the sample quality and is performed on an automated platform. Another advantage of this assay is the provision of genotyping for HPV 16 and 18 [32]. This assay received FDA approval for primary HPV testing without associated Pap cytology in 2014. In April 2020, automated COBAS 6800/8800 systems also received USFDA approval [33].

The Onclarity HPV test (Becton Dickinson) targets the E6/E7 coding region of the HPV viral genome and delineates HPV 16, 18, 31, 45, 51, 52 individually and 33/58, 56/59/66, and 35/39/68 in groups. This kit received FDA approval for primary HPV testing in 2018 [34]. The COBAS HPV real-time PCR test is also approved by FDA for co-testing. Apart from this, there are six more kits that the FDA has approved for co-testing.

The first HPV test for cervical sample screening approved by FDA was the digene hybrid capture 2 HPV DNA test (HC 2 Qiagen). It is a signal amplification-based test that uses RNA probes to detect HPV DNA through DNA:RNA hybridization. During signal amplification, the DNA:RNA complexes are detected using alkaline phosphatase conjugated with multiple antibodies, and the signal generated by the chemiluminescent reaction is read and interpreted automatically. Though capable of detecting 13 high-risk HPV types (16, 18, 31, 33, 35, 39, 45, 51, 52, 56, 58, 59, and 68), this kit is not capable of discriminating between the genotypes and may lead to false-positive results due to a cross-reaction with some of the nononcogenic types. This assay does not assess the quality of the sample by detecting the presence of human cells in the sample. An automated version called rapid capture system is available for throughput testing of a high-volume sample [35].

The Cervista HPV HR assay (Hologic, Marlborough, MA) is another signal amplification-based FDA-approved assay, which utilizes the principle of Invader chemistry and hybridization. Compared to the digene hybrid capture 2 HPV DNA test, it detects an additional high-risk genotype, viz. HPV 66, and contains control probes against histones as a quality control mechanism to confirm the adequacy of the samples. Like the digene hybrid capture 2 HPV DNA assay, this test cannot differentiate between the genotypes and can lead to false-positive results with cross-reacting genotypes. The assay proceeds through two isothermal reactions, primary and secondary. In the primary reaction, two types of HPV E6/E7 sequence-specific oligonucleotides (i.e., an invader oligonucleotide and a probe nucleotide) bind to the target HPV sequence. When such oligonucleotides overlap on the target sequence by at least one base pair, an invasive structure is formed that works as a substrate for the Cleavase enzyme. The enzyme cleaves the 5′ end of the probe at the overlapping site, which then triggers the secondary reaction. In this secondary reaction, it is like an invader probe for the universal fluorescence resonance energy transfer (FRET) signal oligonucleotide. This again activates the Cleavase enzyme to remove the fluorophore from its quencher on the universal oligonucleotide. A genotyping assay is available as a reflex test from the same manufacturer, termed Cervista HPV 16/18, that differentiates between HPV types 16 and 18 [36].

Aptima HPV assay (Hologic) depends on the detection of the same HPV targets E6/E7 and the same genotypes as Cervista HPV HR assay but employs the principle of transcription-mediated amplification to amplify E6/E7 mRNA copies. The working principle includes a single-tube reaction including target capture using target-specific oligomers bound to magnetic microparticles, target amplification by transcription-mediated amplification, and the detection of the amplicons by the hybridization protection assay (HPA) using single-stranded oligonucleotide probes with chemiluminescent labels complementary with the target sequence. A separate genotype assay is also available as Aptima HPV 16, 18/45 which incorporates genotype detection [23,37].

While these assays employ complex laboratory protocols, the need for alternative tests that can be undertaken in low- and middle-income countries has led to the development of point- of-care assays like Xpert HPV (Cepheid) and *Care*HPV (Qiagen). In fact, in Dec 2017, Xpert HPV was the first test approved for the WHO list of prequalified in vitro diagnostics for CC screening. It is a multiplex real-time PCR reaction targeting the E6/E7 region of viral DNA across 14 high HR-HPVs in a single reaction (types HPV 16 and HPV 18/45 in two different detection channels, and 11 other HR-HPVs viz. 31, 33, 35, 39, 51, 52, 56, 58, 59, 66, and 68 in a pooled result). Specimens must be collected in PreservCyt solution (Hologic Corp) [38].

3. Screening strategies

Screening strategies have been extremely effective in decreasing the CC incidence in the developed world. While the age-standardized incidence rate for CC has been estimated to be 10/100,000 in resource-rich countries, the same stands at 25–55/100,000 in resource-poor countries [39]. The rationale for strengthening screening for CC is also underscored by the fact that an overwhelming 85% of new cases are reported from developing countries that are yet to implement successful screening programs [40]. Apart from the population coverage of screening programs, the screening techniques also differ in resource-rich and resource-poor settings. But despite these differences, the underlying goal of all screening techniques involves the detection of treatable precancerous lesions like adenocarcinoma, CIN2, and CIN3 in situ and thus reduces the occurrence of CC, resulting in mortality and treatment-associated morbidity. In addition to the detection of precursor lesions, identification of early-stage invasive cancer through systematic screening can also yield similar benefits. Alongside the requirement of adequate sensitivity in detecting treatable early premalignant lesions and early-stage cancers, screening strategies also need to be sufficiently specific to avoid overtreatment of transient HPV infections and associated benign lesions [41].

In resource-rich settings, as illustrated in the 2020 recommendations of the American Cancer Society (ACS), CC screening is initiated at the age of 25 and is preferably undertaken through primary HPV testing by an FDA-approved test every 5 years. In the case of unavailability of primary HPV testing, co-testing for HPV and cytology is recommended at 5-year intervals or cytology alone at 3-year intervals is also considered acceptable [42]. This is a significant shift from the 2012 recommendations of ACS, wherein cytology alone at 3-year intervals was the modality advocated for women between 21 and 29 years, and co-testing every 5 years was the preferable modality (3-yearly cytology being an acceptable alternative) in the age-group of 30–65 years [43]. Though cytologic screening had been the mainstay of screening for decades, methods based on cytology or co-testing are likely to be removed from future guidelines in view of the trend of transition to HPV nucleic acid-based screening tests. Currently, the utilization of co-testing has been reported to have increased while reliance on cytology alone has fallen [44–48]. Beginning at the age of 25, the screening tests can be discontinued at the age of 65 years in individuals who had not reported CIN grade 2 or above in the past 25 years and people who have had acceptable negative screening results in the past 10 years. The criterion for adequate negative screening results is said to be fulfilled when two consecutive HPV tests or co-tests or three consecutive cytology tests are reported negative, with the most recent test being performed within the recommended interval for that test. Individuals with abnormal screening results are required to continue screening even after 65 years of age, provided they do not suffer from conditions associated with limited life expectancy

[42]. Management of abnormal screening results is guided according to the 2019 recommendations of the American Society for Colposcopy and Cervical Pathology (ASCCP) for the management of screening abnormalities, which include modification of the frequency of surveillance, colposcopy, treatment, or return to routine surveillance [49]. While screening can be discontinued after hysterectomy, vaccination for HPV is not reviewed to be a ground for discontinuation of screening. It is worth noting that these screening recommendations apply to individuals with average risk and do not apply to individuals with a higher risk of developing CC, e.g., people living with HIV/AIDS, transplant recipients, patients on immunosuppressant therapy, and those exposed to diethylstilbestrol in utero. For example, according to the Centers for Disease Control and Prevention (CDC) recommendations, in HIV seropositive individuals below the age of 30 years, screening by cytology is suggested within the former year of diagnosis and thereafter at annual intervals till three normal test results are obtained. For HIV-positive women above the age of 30 years, co-testing can be an alternative screening modality [50].

In contrast to the situation in developed countries, economic and logistic obstacles impede the implementation of successful screening programs in developing countries. In 2016, the Government of India initiated a program for a population-based screening of females within the age-group of 30–65 years using visual inspection with acetic acid (VIA) every 5 years [51]. Once found to be positive on VIA, the individual is referred for expert consultation and eligible lesions are treated with cryotherapy and follow-up after 1 year. Only if the lesion is ineligible for cryotherapy, a naked eye/colposcopy-guided biopsy is taken for histopathological grading and this guides further management. A CIN 1 lesion is treated with cryotherapy and followed up with VIA after 1 year while CIN 2 and 3 lesions are cured with a loop electrosurgical excision procedure (LEEP). This is in accordance with the WHO guidelines [52]. In Indonesia, VIA-based tests are used to screen and treat married women of the age-group 30–50 years since 2014 [53]. In Thailand, 5-yearly screening based on VIA and cytology has been used since 2005 in a national program targeting women between 35 and 60 years. Recently, there has been a shift to introduce HPV screening in the primary screening armamentarium in Thailand [53]. In Africa, especially in the sub-Saharan Africa (SSA) region, several initiatives are being taken to improve uptake for a population-based screening [54]. In Kenya, the national guidelines of 2018 recommend screening for females of the age-group of 25–49 years. HPV testing is considered the primary choice for those above 30 years and VIA with or without visual examination under Lugol's iodine (VILI) can also be used where HPV testing is unavailable or in individuals < 30 years of age. Use of Pap smear cytology is limited to specified circumstances like nonavailability of HPV testing, ineligibility for VIA/VILI, or as a co-test with HPV testing in HIV-positive women [55].

4. Conclusion

Despite rapid evolution in screening strategies and substantial reduction in invasive CC frequency especially in the developed world, there remains a need for active research in several areas to create evidence-based strategies for effective and universal screening programs. Specifically, obstacles impeding the successful implementation of population-centered screening programs in the low- and middle-income countries (LMIC) need to be navigated to yield translational benefits of emerging technologies across the globe. The development of cost-effective point-of-care devices for regions with inadequate laboratory infrastructure along with the integration of self-collected samples in screening guidelines can go a long way in improving the population coverage of screening. Undertaking economic evaluation of different strategies by a comparative analysis of cost-benefit ratio may aid developing countries in selecting the appropriate public health strategy for containing CC. The assessment of risk-stratified screening protocols, compared to universal screening strategies, could be explored in randomized controlled trials in countries with limited resources. With expanding immunization, the need for tailoring screening strategies for the vaccinated populations should be evaluated. Given the role of HPV in glandular neoplasms of the cervix and noncervical neoplasms, the scope of screening strategies needs to be broadened to encompass these categories as well.

References

[1] HPV-Associated Cancer Statistics | CDC [Internet], 2020. [cited 2021 Oct 4]. Available from: https://www.cdc.gov/cancer/hpv/statistics/index.htm.

[2] C. de Martel, D. Georges, F. Bray, J. Ferlay, G.M. Clifford, Global burden of cancer attributable to infections in 2018: a worldwide incidence analysis, Lancet Glob. Health 8 (2) (2020) e180–e190.

[3] P.M. Howley, J.T. Schiller, D.R. Lowy, Papillomaviruses, in: D.M. Knipe, P.M. Howley (Eds.), Fields Virology, sixth ed., vol. 1, Lippincott Williams & Wilkins, a Wolters Kluwer, Philadelphia, PA, 2013, pp. 1662–1703.

[4] W. Bonnez, R.C. Reichman, Papillomaviruses, in: G.L. Mandell, J.E. Bennett, R. Dolin (Eds.), Mandell, Douglas, and Bennett's Principles and Practice of Infectious Diseases, fifth ed., Churchill Livingston, Philadelphia, PA, 2000, pp. 1630–1640.

[5] F. Bosch, M.M. Manos, N. Munoz, M. Sherman, A.M. Jansen, J. Peto, M.H. Schiffman, V. Moreno, R. Kurman, K.V. Shah, An international biological study on cervical cancer (IBSCC) study Group. Prevalence of human papillomavirus in cervical cancer: a worldwide perspective, J. Natl. Cancer Inst. 87 (1995) 796–802.
[6] J. Doorbar, N. Egawa, H. Griffin, C. Kranjec, I. Murakami, Human papillomavirus molecular biology and disease association, Rev. Med. Virol. 25 (Suppl 1) (2015) 2–23. Mar [cited 2021 Sep 29]. Available from: https://www.ncbi.nlm.nih.gov/pmc/articles/PMC5024016/.
[7] World Health Organization, Human Papillomavirus Laboratory Manual [Internet], World Health Organization, 2010. [cited 2021 Sep 29]. Report No.: WHO/IVB/10.12. Available from https://apps.who.int/iris/handle/10665/70505.
[8] WHO, Self-Sampling for Human Papillomavirus Testing [Internet], WHO. World Health Organization, 2021. [cited 2021 Sep 29]. Available from: https://www.who.int/teams/sexual-and-reproductive-health-and-research-(srh)/research/journal-articles.
[9] D. Hawkes, M.H.T. Keung, Y. Huang, T.L. McDermott, J. Romano, M. Saville, et al., Self-collection for cervical screening programs: from research to reality, Cancer 12 (4) (2020) 1053. Apr 24 [cited 2021 Sep 27]. Available from https://www.ncbi.nlm.nih.gov/pmc/articles/PMC7226191/.
[10] S. Trevitt, S. Simpson, A. Wood, New and emerging self-sampling technologies for human papillomavirus (HPV) testing, 2014, https://doi.org/10.13140/RG.2.2.12612.63366.
[11] O.A. Ibeanu, Molecular pathogenesis of cervical cancer, Cancer Biol. Ther. 11 (3) (2011) 295–306. Feb [cited 2021 Sep 29]. Available from: http://www.tandfonline.com/doi/abs/10.4161/cbt.11.3.14686.
[12] Definition of Pap Smear—NCI Dictionary of Cancer Terms—National Cancer Institute [Internet], 2011. [cited 2021 Sep 29]. Available from: https://www.cancer.gov/publications/dictionaries/cancer-terms/def/pap-smear.
[13] Definition of Liquid-Based Pap Test—NCI Dictionary of Cancer Terms—National Cancer Institute [Internet], 2011. [cited 2021 Sep 29]. Available from: https://www.cancer.gov/publications/dictionaries/cancer-terms/def/liquid-based-pap-test.
[14] R.K. Gibb, M.G. Martens, The impact of liquid-based cytology in decreasing the incidence of cervical cancer, Rev. Obstet. Gynecol. 4 (Suppl 1) (2011) S2–11.
[15] W. Prendiville, R. Sankaranarayanan, Colposcopy and Treatment of Cervical Precancer [Internet], International Agency for Research on Cancer, IARC Technical Publications, Lyon, 2017. [cited 2021 Oct 21]. Available from: http://www.ncbi.nlm.nih.gov/books/NBK568370/.
[16] J.W. Sellors, R. Sankaranarayanan, Colposcopy and Treatment of Cervical Intraepithelial Neoplasia: A Beginners' Manual, International Agency for Research Cancer, Lyon, 2003. 132 p.
[17] E.M. Burd, C.L. Dean, Human papillomavirus, Microbiol. Spectr. 4 (4) (2016).
[18] Human Papillomavirus (HPV) Self-Sampling as Part of Cervical Cancer Screening [Internet], 2020. [cited 2021 Sep 29]. Available from: https://www.who.int/publications-detail-redirect/WHO-SRH-2012.
[19] M. Arbyn, S.B. Smith, S. Temin, F. Sultana, P. Castle, Detecting cervical precancer and reaching underscreened women by using HPV testing on self samples: updated meta-analyses, BMJ (2018), k4823. [cited 2021 Sep 29];Available from: https://www.bmj.com/lookup/doi/10.1136/bmj.k4823.
[20] R. Nayar, D.C. Wilbur, The Pap Test and Bethesda 2014: "The reports of my demise have been greatly exaggerated. (after a quotation from Mark Twain)", J. Low. Genit. Tract Dis. 19 (3) (2015) 175–184. cited 2021 Sep 29. Available from: https://journals.lww.com/00128360-201507000-00001.
[21] IARC, Cervix Cancer Screening [Internet], 2005, [cited 2021 Sep 29]. Available from: https://publications.iarc.fr/Book-And-Report-Series/Iarc-Handbooks-Of-Cancer-Prevention/Cervix-Cancer-Screening-2005.

[22] R. Nayar, D.C. Wilbur, The Bethesda system for reporting cervical cytology: a historical perspective, Acta Cytol. 61 (4–5) (2017) 359–372. cited 2021 Sep 29. Available from: https://www.karger.com/Article/FullText/477556.
[23] W. Bonnez, Papillomaviruses, in: D.D. Richman, R.J. Whitley, F.G. Hayden (Eds.), Clinical Virology, fourth ed., ASM Press, Washington, DC, 2017, pp. 625–678.
[24] C. PKS, M.A. Picconi, T.H. Cheung, L. Giovannelli, J.S. Park, Laboratory and clinical aspects of human papillomavirus testing, Crit. Rev. Clin. Lab. Sci. 49 (4) (2012) 117–136. cited 2021 Sep 29. Available from: https://www.ncbi.nlm.nih.gov/pmc/articles/PMC3469219/.
[25] C.G. Onyango, L. Ogonda, B. Guyah, C. Shiluli, G. Ganda, O.E. Orang'o, et al., Novel biomarkers with promising benefits for diagnosis of cervical neoplasia: a systematic review, Infect. Agents Cancer 15 (1) (2020) 68.
[26] A.G. Waxman, D. Chelmow, T.M. Darragh, H. Lawson, A.-B. Moscicki, Revised terminology for cervical histopathology and its implications for management of high-grade squamous intraepithelial lesions of the cervix, Obstet. Gynecol. 120 (6) (2012) 1465–1471. cited 2021 Sep 28. Available from: https://www.ncbi.nlm.nih.gov/pmc/articles/PMC4054813/.
[27] Y. Zhu, C. Ren, L. Yang, X. Zhang, L. Liu, Z. Wang, Performance of p16/Ki67 immunostaining, HPV E6/E7 mRNA testing, and HPV DNA assay to detect high-grade cervical dysplasia in women with ASCUS, BMC Cancer 19 (1) (2019) 271.
[28] R. Sankaranarayanan, R. Wesley, S. Thara, N. Dhakad, B. Chandralekha, P. Sebastian, et al., Test characteristics of visual inspection with 4% acetic acid (VIA) and Lugol's iodine (VILI) in cervical cancer screening in Kerala, India, Int. J. Cancer 106 (3) (2003) 404–408.
[29] A. Arney, K.M. Bennett, Molecular diagnostics of human papillomavirus, Lab. Med. 41 (9) (2010) 523–530. cited 2021 Oct 22. Available from https://doi.org/10.1309/LM75WGJSVMI7VVEF.
[30] A. Gradíssimo, R.D. Burk, Molecular tests potentially improving HPV screening and genotyping for cervical cancer prevention, Expert Rev. Mol. Diagn. 17 (4) (2017) 379–391. cited 2021 Oct 22. Available from: https://www.ncbi.nlm.nih.gov/pmc/articles/PMC5904788/.
[31] M. Poljak, B.J. Kocjan, A. Oštrbenk, K. Seme, Commercially available molecular tests for human papillomaviruses (HPV): 2015 update, J. Clin. Virol. 76 (Suppl 1) (2016) S3–13.
[32] cobas® 4800 System [Internet], 2021. [cited 2021 Oct 22]. Available from: https://diagnostics.roche.com/global/en/products/systems/cobas-4800-system.html.
[33] Health C for D and R, The Cobas HPV for Use on the cobas 6800/8800 Systems—P190028. FDA [Internet], 2020, [cited 2021 Sep 29]; Available from: https://www.fda.gov/medical-devices/recently-approved-devices/cobas-hpv-use-cobas-68008800-systems-p190028.
[34] BD Onclarity™ HPV Assay [Internet], 2021. [cited 2021 Oct 4]. Available from: https://www.bd.com/en-us/offerings/capabilities/cervical-cancer-screening/onclarity-hpv-assay.
[35] Digene HC2 High-Risk HPV DNA Test [Internet], 2021. [cited 2021 Oct 4]. Available from: https://www.qiagen.com/us/products/diagnostics-and-clinical-research/sexual-reproductive-health/cervical-cancer-screening/digene-hc2-high-risk-hpv-dna-test/.
[36] Cervista HPV HR.pdf [Internet], 2021. [cited 2021 Oct 4]. Available from: https://www.hologic.com/sites/default/files/package-insert/15-3100_105_01.pdf.
[37] APTIMA HPV ASSAY.pdf [Internet], 2021. [cited 2021 Oct 4]. Available from: https://www.hologic.com/sites/default/files/package-insert/AW-14517-001_003_01.pdf.
[38] 171221_final_pq_report_pqdx_0268_070_00.pdf [Internet], 2017. [cited 2021 Sep 29]. Available from: https://extranet.who.int/pqweb/sites/default/files/PQDx0268-070-00-XpertHPV_4.0.pdf.
[39] IARC Publications Website—Cancer Incidence in Five Continents Volume VIII [Internet], 2002. [cited 2021 Sep 29]. Available from: https://publications.iarc.fr/Book-And-Report-Series/Iarc-Scientific-Publications/Cancer-Incidence-In-Five-Continents-Volume-VIII-2002.

[40] H. Sung, J. Ferlay, R.L. Siegel, M. Laversanne, I. Soerjomataram, A. Jemal, et al., Global Cancer statistics 2020: GLOBOCAN estimates of incidence and mortality worldwide for 36 cancers in 185 countries, CA Cancer J. Clin. 71 (3) (2021) 209–249.
[41] New Recommendations for Screening and Treatment to Prevent Cervical Cancer [Internet], 2021. [cited 2021 Oct 22]. Available from: https://www.who.int/news/item/06-07-2021-new-recommendations-for-screening-and-treatment-to-prevent-cervical-cancer.
[42] E.T. Fontham, A.M. Wolf, T.R. Church, R. Etzioni, C.R. Flowers, A. Herzig, C.E. Guerra, K.C. Oeffinger, Y.C. Shih, L.C. Walter, J.J. Kim, Cervical cancer screening for individuals at average risk: 2020 guideline update from the American Cancer Society, CA Cancer J. Clin. (2020). Wiley Online Library [Internet]. [cited 2021 Sep 29]. Available from: https://acsjournals.onlinelibrary.wiley.com/doi/10.3322/caac.21628.
[43] D. Saslow, D. Solomon, H.W. Lawson, M. Killackey, S.L. Kulasingam, J. Cain, et al., American Cancer Society, American Society for Colposcopy and Cervical Pathology, and American Society for Clinical Pathology screening guidelines for the prevention and early detection of cervical cancer, CA Cancer J. Clin. 62 (3) (2012) 147–172.
[44] US Preventive Services Task Force, S.J. Curry, A.H. Krist, D.K. Owens, M.J. Barry, A.B. Caughey, et al., Screening for cervical cancer: US preventive services task force recommendation statement, JAMA 320 (7) (2018) 674–686.
[45] C. Zhao, B.A. Crothers, M. Ghofrani, Z. Li, R.J. Souers, M. Hussain, et al., Human papillomavirus genotyping testing practice in 2014: results of a College of American Pathologists national survey, Arch. Pathol. Lab. Med. 140 (12) (2016) 1364–1370.
[46] C. Zhao, A.T. Moriarty, M. Ghofrani, M. Husain, R.H. Tambouret, R. Laucirica, et al., Human papillomavirus testing and reporting rates in 2012: results of a College of American Pathologists national survey, Arch. Pathol. Lab. Med. 139 (6) (2015) 757–761.
[47] M. Watson, V. Benard, E.W. Flagg, Assessment of trends in cervical cancer screening rates using healthcare claims data: United States, 2003-2014, Prev. Med. Rep. 9 (2018) 124–130.
[48] M. Watson, V. Benard, J. King, A. Crawford, M. Saraiya, National assessment of HPV and Pap tests: changes in cervical cancer screening, National Health Interview Survey, Prev. Med. 100 (2017) 243–247.
[49] 2019 ASCCP Risk-Based Management Consensus Guidelines for Abnormal Cervical Cancer Screening Tests and Cancer Precursors [Internet], 2020. [cited 2021 Sep 29]. Available from https://www.ncbi.nlm.nih.gov/pmc/articles/PMC7147428/.
[50] Human Papillomavirus Disease | NIH [Internet], 2021. [cited 2021 Sep 29]. Available from: https://clinicalinfo.hiv.gov/en/guidelines/adult-and-adolescent-opportunistic-infection/human-papillomavirus-disease.
[51] Operational_Framework_Management_of_Common_Cancers.pdf [Internet], 2016. [cited 2021 Sep 29]. Available from: https://main.mohfw.gov.in/sites/default/files/Operational%20Framework%20Management%20of%20Common%20Cancers_1.pdf.
[52] WHO Guideline for Screening and Treatment of Cervical Pre-Cancer Lesions for Cervical Cancer Prevention [Internet], 2021. [cited 2021 Sep 29]. Available from: https://www.who.int/publications/i/item/9789240030824.
[53] E.S. Aoki, R. Yin, K. Li, N. Bhatla, S. Singhal, D. Ocviyanti, et al., National screening programs for cervical cancer in Asian countries, J. Gynecol. Oncol. 31 (3) (2020) e55. cited 2021 Oct 1. Available from https://ejgo.org/DOIx.php?id=10.3802/jgo.2020.31.e55.
[54] N.B. Yimer, M.A. Mohammed, K. Solomon, M. Tadese, H.K. Meikena, B. Alemnew, et al., Cervical cancer screening uptake in Sub-Saharan Africa: a systematic review and meta-analysis, Public Health (2021) 25.
[55] National-Cancer-Screening-Guidelines-2018.pdf [Internet], 2018. [cited 2021 Oct 1]. Available from: https://www.health.go.ke/wp-content/uploads/2019/02/National-Cancer-Screening-Guidelines-2018.pdf.

CHAPTER 7

Animal models for the discovery of therapeutic and prophylactic measure against HPV-induced malignancies

Shobha Upreti and Mukesh Samant
Cell and Molecular Biology Laboratory, Department of Zoology, Kumaun University, Almora, Uttarakhand, India

1. Introduction

The human papillomaviruses (HPVs) consist of a nonenveloped, small, circular dsDNA, with an approximate size of 7.9–8 kbp [1]. These viruses are surrounded by a tetrahedral capsid and consist mainly of the L1 protein which is highly immunogenic, along with a less immunogenic L2 protein [2,3]. The virus induces proliferation of the squamous epithelial cells, or the formation of common warts, on the feet, hands, anus, scrotum, cervix, thighs, groin, or penis [4]. There are five stages in the life cycle of HPV. The first phase of the life cycle includes infection and simultaneous uncoating, followed by the second stage that involves genome maintenance, which includes the expression of E1 and E2, i.e., the early proteins. The third stage is the proliferative phase, which includes the expression of E6 and E7, along with some other early proteins, which are crucial for the progression of the cell cycle. The proliferative phase is then followed by the fourth stage where the L1 and L2, i.e., the late proteins, are expressed. The expression of the L1 and L2 proteins leads to the packaging of the virus inside the epithelial layer. Furthermore, the virus is released in the fifth phase, consequently infecting the other cells [5,6].

HPVs are responsible for the induction of various dermatological and sexually transmitted diseases; moreover, over 130 HPVs have been reported to cause several malignancies in humans [1]. Based on the occurrence of different lesions, HPV genotypes have been classified into two groups, i.e., the low-risk HPV genotypes (HPV 6 and HPV 11) that are responsible for the benign lesions, viz., cancers, recurrent respiratory papillomatosis (RRP), genital warts, etc. [7] and the high-risk HPV (HR-HPV) genotypes (HPV 16, HPV 18, HPV 31, HPV 33, and HPV 5) that are responsible for almost all types of cervical cancers [8].

Immunopathology, Diagnosis, and Treatment of HPV-induced Malignancies
https://doi.org/10.1016/B978-0-323-91797-1.00008-9

Cervical cancer is one of the leading causes of cancers in women worldwide; hence, it has been reported to cause a considerable number of deaths [1]. This type of cancer is caused by the HR-HPVs, which include more than 100 strains of HPV [1,9]. HPV has also been classified into different types based on the geographical regions and histological types; for example, different strains of HPVs have been reported based on the histological and geographical differences in about 30,848 cases of invasive cervical cancers worldwide [10]. Infections associated with HPV are very common; hence, HPV vaccination that can prevent genital warts, precancerous lesions, and cervical cancers in young individuals has been a public health priority all around the globe. Such vaccination strategies can progressively decrease the number of HPV cases [11]. Therefore, animal models are crucial in developing prophylactic and therapeutic measures against HPV; moreover, these model systems allow researchers to examine the entire life cycle of the papillomavirus (PV) [8]. Besides, the immunological response of the host to the natural papillomavirus infection can also be studied by using particular animal models [1,8]. Therefore, the present chapter deals with the best-studied animal models of HPV that include rabbits, dogs, cattle, along with the more recent model systems, i.e., the mouse papillomavirus (MmuPV) studied in nude mice, and transgenic animals (Fig. 1).

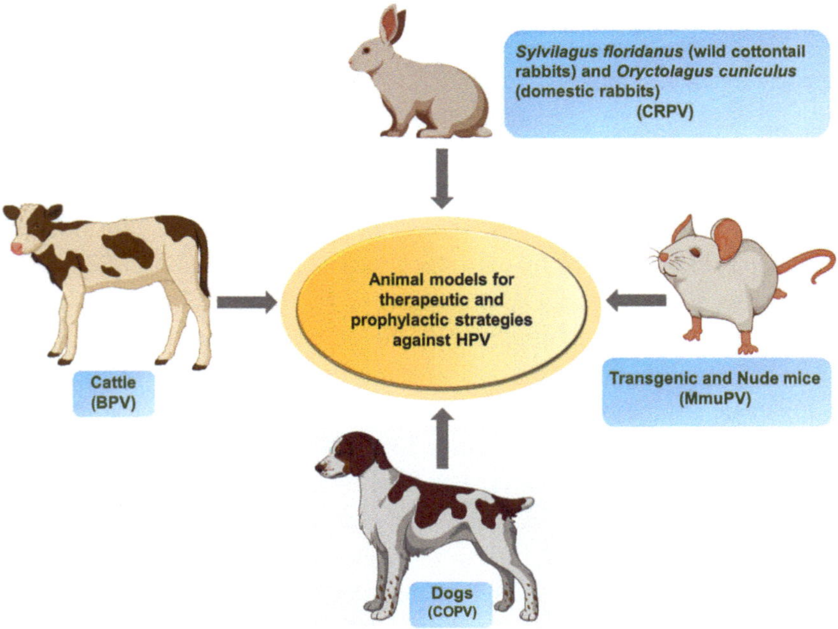

Fig. 1 Animal models involved in the development of various therapeutic and prophylactic strategies against HPV.

2. Animal models for therapeutic and prophylactic measures against HPV

Animal models that can mimic the malignancies caused by HPV in a similar experimental system are crucial to combat HPV. Moreover, the efficiency and optimization of the prophylactic and therapeutic approaches can also be studied in the animal model system before the occurrence of clinical trials.

2.1 *Sylvilagus floridanus* (wild cottontail rabbits) and *Oryctolagus cuniculus* (domestic rabbits)

The first tumor-causing DNA virus to be characterized was the cottontail rabbit papillomavirus (CRPV) [12]. In natural and experimental settings, CRPV has been reported to be the causative agent of cutaneous papillomas in cottontail rabbits [12]. The natural host of CRPV is *Sylvilagus floridanus* (wild cottontail rabbits), but this virus can also infect, a very similar species, i.e., the *Oryctolagus cuniculus* (domestic rabbits) [13]. Since the wild cottontail rabbits possess a range of genetic variations and are difficult to be caged, the domestic rabbit has been widely used for experimental purposes. Although when white rabbits in New Zealand were studied for the lesions caused by CRPV, it was reported that the productive life cycle was not completed in them, resulting in the production of hardly any virions [12].

The domestic rabbit-CRPV model gives highly reproducible and consistent results and can be managed easily [8]. Rabbits vaccinated with the L1 or L2 proteins induced protection against the CRPV infection, but comparatively, the L1 protein was reported to exhibit robust immune response [8]. Moreover, rabbits vaccinated with these structural proteins were also reported to display the production of neutralizing antibodies against this infection [8]. In a considerable number of rabbits, specific T cell-mediated immunity has been observed to play a vital role in regressing the CRPV-induced lesions, consequently limiting tumor growth. Hence, the regressing lesions indicate the active functioning of the host immune system against carcinogenic capabilities of CRPV similar to that of HPV [8,14]. This makes the rabbit-CRPV model an excellent model system to test a range of prophylactic and therapeutic approaches that can also be applied to the HPVs. However, CRPV exclusively infects hairy skin organisms, therefore making it an incompetent model for mucosa-tropic HPV. In addition, CRPVs are unable to complete their life cycle in their experimental host, i.e., the domestic rabbits; hence, an alternative host which can overcome this limitation is required [12].

2.2 Cattle

Another animal model involved in the development of effective anti-HPV vaccination strategies is the bovine papillomavirus (BPV) model. The majority of BPVs have been reported to infect the cutaneous epithelium of cattle [12]. In comparison to the other PVs, BPV can infect a diverse range of fibroblast cells in a large number of ungulates [12]. Moreover, it can also cause various diseases in cattle, viz., skin warts, urinary bladder cancer, upper gastrointestinal tract cancer, etc. [13] BPV-4, on the other hand, has been reported to infect the mucosal epithelium of the proximal alimentary canal, causing squamous cell carcinomas [15]. Hence, BPV can infect the skin or the mucosa similar to HPVs, sometimes leading to cancerous conditions.

The prophylactic and therapeutic immunization in calves has been reported to involve the L1 and L2 PV proteins [8]. Vaccination with the L1 protein was reported to induce prophylactic protection in them, whereas immunization with the L2 protein protected calves irrespective of the time of its administration. The calves immunized with the L1 protein induced neutralizing antibody production, in contrast to the calves vaccinated with the L2 protein, which promoted the production of cytotoxic T cells. Moreover, when calves were vaccinated with the N terminus of the L2 protein, they were found to be fully protected from this infection by producing enough amount of neutralizing antibodies [8]. Unlike CRPV, where L1 immunization was most effective, in BPV, immunization with the L2 protein has been reported to be more effective, demonstrating that different immunization strategies can be effective in different model systems. However, as the L1 and L2 proteins are not expressed in HPVs that cause human cancers, the vaccines containing E7 protein are optimal for vaccination against HPV. Therefore, to measure the efficacy of the E7-based vaccine strategies, the BPV E7 protein was administered to the calves, and cell-mediated immune response was measured [8]. The vaccine constituting the E7 protein resulted in the slowed growth of the papillomas and speedy recovery of the lesions [8]. Moreover, infectious BPV can be procured in huge quantities from the warts developed in infected cattle, due to the presence of the host tissue or the bovine skin in ample amounts [16]. Consequently, this model system resolves the most basic concerns, viz., tissue availability and variability, of using the human tissues for research purposes [8,12]. However, providing shelter to such huge animals is a challenge for researchers [8,12].

2.3 Dogs

Canine oral papillomavirus (COPV) promotes wart formation in the oral mucosa of the wild canines and domestic dogs, analogous to that caused due to low-risk HPVs. Merely, some of the PVs can induce the development of higher-grade lesions like the oral squamous cell cancers, compared to the low-grade lesions which produce a large number of transmissible viruses, that are easy to isolate. [8]. Both human and dog cells have a lot in common, such as the antigen markers present on their cell surfaces, along with the functioning of the antigen-presenting cells, and the lymphocytes [17]. These similarities, make dogs a suitable model system to develop low-risk HPV vaccines. Lesions due to COPV infection take 4 to 8 weeks to develop, accompanied by the immune-mediated regression in another 4–8 weeks [12]. Analogous to the papilloma associated with HPV, wart regression results in the invasion of the lymphocytes inside the papillomas; subsequently, these lymphocytes reach their peak throughout the resolution of the infection and then again decline to their initial levels once the lesion has been cleared [17,18]. Dogs with spontaneous regressing papillomas can be studied to isolate the COPV-specific neutralizing IgG antibodies [19]. Moreover, the passive serum transfer from the host with regressing lesions can help to prevent the COPV-induced papillomas, suggesting that these IgG antibodies play an important role in preventing the intraoral infection caused by the COPV. Immunocompromised animals develop serious illness, similar to those infected with HPV [17]. COPV have been extensively used for research purpose; however, the COPV model has been replaced in recent years with advancements in the other model systems.

2.4 *Mus musculus* (nude mice)

A PV was discovered in the common house mouse, *Mus musculus*, in 2010 from the spontaneous lesions developed in the NMRI-Foxn1(nu)/Foxn1(nu) nude mice colony [20]. Even though the cloned MmuPV genome was not provided to the PV scientific community initially, MmuPV was chemically synthesized, once its whole genome sequence was published [21], and this type of HPV has been used in numerous investigations. Since then, many researchers have independently extracted MmuPV from the mouse tissues [22]. However, like all the animal PVs, an evolutionary difference exists between the animal PVs and HPVs, which suggests that the model systems are unable to display certain characteristics of HPV; instead, these may provide general insights into how PVs influence the immune system and epithelial cells they colonize [12]. Interestingly, MmuPV, a member of the πPV family, appears to induce evident papillomas in the immunocompromised mice, exclusively; however, the lesion formation was found to be suppressed in immunocompetent mice [12]. Thus, MmuPv infections seem to be closely related to the infection caused by β-HPVs and have been reported to cause transmissible undetected infections. These papillomas can be visible in the immunocompromised model [23]. Such visible papillomas that represent productive infections for both cases involve significant genome replication along with a delayed gene expression [24,25].

Interestingly, the tropism for skin infections was primarily determined for MmuPV and βHPV, but recent studies have proved that β and γ forms of HPV are present in mucosal surfaces along with the oral cavity [26], indicating that MmuPV can cause infection in the genital mucosal tissue as well [27]; hence, it was observed to infect a broader tissue range than was initially thought [28]. However, due to such a broad tissue range and association with cervical lesions, this particular model has been reported to provide information about human αPV in the cervical region [27,29]. Moreover, this model, in addition to providing general information about how PVs interact with both epithelial cells and the immune system, also provides insights into how the human β-PVs and γ-PVs can induce different infections [30,31].

2.5 Transgenic animals

Recently, the models that can be infected with the animal papillomaviruses in their natural hosts have been proven to be quite useful; however, these models are unable to address the precise actions and maladies caused in people after infection with certain types of HPV [1,12]. The PVs and their respective hosts have coevolved along with a limited amount of interspecies transmission; hence, a range of pathological correlations and molecular mechanisms are not substantially common among the divergent PVs [32–34]. This issue has been somehow resolved by using an animal PV transgenic model system. Transgenic mice have been developed for the constitutive expression of the viral proteins under the control of an exogenous promoter [35–38]. The expression of the E6 and E7 early proteins of HPV16 regulated by the basal cell K14 (keratin 14) promoter has been studied in several tissues [35,36,38]. These investigations imply that E6 is responsible for the malignant phenotypes; moreover, the effect of viral genes varies with the change in the epithelial sites, which can be modified by the external stimuli, viz., the hormonal signaling or the cytokines [35,36,38].

Hence, this model has the potential to understand the events leading to cancer, as the malignancies caused by HPV16 are often associated with the loss in the expression of the E2 early protein and constitutive expression of the E6 and E7 early proteins [12]; hence, this model yields valuable information regarding HPV16 along with other types of HPV. Moreover, studies have been conducted where the K14 promoter was utilized for the expression of the HPV8 gene in transgenic mice [39] and was reported that E6 and E2 are crucial for the formation of malignant skin tumors [40,41]; besides, UV irradiation substantially accelerated this process. The limitation to this model system is that a k14 promoter was present to regulate the gene expression instead of an endogenous promoter of PV, which is responsible for the examination of the effect of estrogen-like hormones on viral transcription [12]. Moreover, transgenic models are dependent on the viral proteins that are developed in humans and act similarly in the model organism. However, the presence of the protein-protein interaction in PVs to induce different actions causes variations in the functioning of the HPV proteins in the mouse cells [12].

3. Conclusion

Animal PVs are crucial as models for HPV infections. For several years, BPV and CRPV have been used as the model systems for studying HPV biology. Animal models like cattle and rabbits have been used to demonstrate and reproduce the experiments regarding the induction and progression of the papillomas. The study of HPV has been less challenging since the advances in cell culture and molecular biology techniques. Moreover, further study of the animal PV, including the similarity between the BPV and HPVs involved in the bladder and esophageal cancers, along with the site of viral latency may provide insights into the viral research [42]. Reports suggest that lymphocytes have been the site of viral latency for both HPV and BPV, hence making the animal model system a crucial part of HPV research [42]. Cattle, rabbits, and more recently dogs have provided opportunities to investigate vaccination in the natural host, resulting in the development of effective vaccines. Based on such discoveries, several preventative and therapeutic anti-HPV vaccines that are presently in the clinical trials with encouraging results have been developed [1,42]. Moreover, despite a wide range of models are available to study HPV-related diseases, the advantages and limitations of various models are considered carefully. It is crucial to know how effectively a specific model simulates the epithelial site to be studied; hence, to do so, a thorough understanding of the clinical illness and patterns related to the expression of the viral gene at distinct epithelial sites are observed. Another challenge is that the HR-HPV and LR-HPVs certainly have functional variations in their expressed proteins as well in the organization of respective life cycles [12]. Hence, both the HPVs cannot be studied in the same model unless these challenges are taken into consideration. Infection by HPV can be directly evaluated through animal models, which allows PVs to develop in their natural hosts, consequently, permitting the analysis of viral latency and immune responses against HPV-induced lesions [12]. This can be significantly improved by using the animal model where the productive life cycle of PV can be completed. Even though these areas play a crucial role in PV studies, there are considerable evolutionary gaps among various animal PVs and clinically significant HPVs. Moreover, only the PV and host interaction can be studied in a few cases instead of the information about a particular type of HPV. Hence, such cases require specific model systems for the observational studies of the HPV-related clinical illness, along with the molecular studies that examine protein function and regulation of the expression of viral genes. HPV disease models will continue to be crucial in increasing our knowledge of viral biology, and when employed as part of the multidisciplinary strategy, they will contribute toward the emergence of effective therapies to prevent the occurrence of such diseases.

References

[1] M. Dadar, S. Chakraborty, K. Dhama, et al., Advances in designing and developing vaccines, drugs and therapeutic approaches to counter human papilloma virus, Front. Immunol. 9 (2018) 2478.
[2] I.H. Frazer, Prevention of cervical cancer through papillomavirus vaccination, Nat. Rev. Immunol. 4 (2004) 46–55.
[3] G.R. Leggatt, I.H. Frazer, HPV vaccines: the beginning of the end for cervical cancer, Curr. Opin. Immunol. 19 (2007) 232–238.
[4] M.A. Stanley, Epithelial cell responses to infection with human papillomavirus, Clin. Microbiol. Rev. 25 (2012) 215–222.
[5] M. Stanley, Pathology and epidemiology of HPV infection in females, Gynecol. Oncol. 117 (2010) S5–10.
[6] D.E. LaCour, C. Trimble, Human papillomavirus in infants: transmission, prevalence, and persistence, J. Pediatr. Adolesc. Gynecol. 25 (2012) 93–97.
[7] M. Stanley, Immunobiology of HPV and HPV vaccines, Gynecol. Oncol. 109 (2008) S15–S21.
[8] S.C. Fausch, D.M. da Silva, G.L. Eiben, et al., HPV protein/peptide vaccines: from animal models to clinical trials, Front. Biosci. 8 (2003) 81–91.
[9] H. Zur Hausen, Papillomaviruses and cancer: from basic studies to clinical application, Nat. Rev. Cancer 2 (2002) 342–350.

[10] N. Li, S. Franceschi, R. Howell-Jones, et al., Human papillomavirus type distribution in 30,848 invasive cervical cancers worldwide: variation by geographical region, histological type and year of publication, Int. J. Cancer 128 (2011) 927–935.
[11] P. Bonanni, A. Bechini, R. Donato, et al., Human papilloma virus vaccination: impact and recommendations across the world, Ther Adv Vaccines 3 (2015) 3–12.
[12] J. Doorbar, Model systems of human papillomavirus-associated disease, J. Pathol. 238 (2016) 166–179.
[13] S.I. Ruiz, E.E. Zumbrun, A. Nalca, Animal models of human viral diseases, in: Animal Models for the Study of Human Disease, 2017, p. 853.
[14] G. Knowles, B.W. O'Neil, M.S. Campo, Phenotypical characterization of lymphocytes infiltrating regressing papillomas, J. Virol. 70 (1996) 8451–8458.
[15] F.W. Hill, W.R. Klein, M.J. Hoyer, et al., Antitumor effect of locally injected low doses of recombinant human interleukin-2 in bovine vulval papilloma and carcinoma, Vet. Immunol. Immunopathol. 41 (1994) 19–29.
[16] A. Pawellek, G. Hewlett, J. Kreuter, et al., Xenograft model for identifying chemotherapeutic agents against papillomaviruses, Antimicrob. Agents Chemother. 45 (2001) 1014–1021.
[17] P.K. Nicholls, B.A. Klaunberg, R.A. Moore, et al., Naturally occurring, nonregressing canine oral papillomavirus infection: host immunity, virus characterization, and experimental infection, Virology 265 (1999) 365–374.
[18] P.K. Nicholls, P.F. Moore, D.M. Anderson, et al., Regression of canine oral papillomas is associated with infiltration of CD4+ and CD8+ lymphocytes, Virology 283 (2001) 31–39.
[19] S. Ghim, J. Newsome, J. Bell, et al., Spontaneously regressing oral papillomas induce systemic antibodies that neutralize canine oral papillomavirus, Exp. Mol. Pathol. 68 (2000) 147–151.
[20] A. Ingle, S. Ghim, J. Joh, et al., Novel laboratory mouse papillomavirus (MusPV) infection, Vet. Pathol. 48 (2011) 500–505.
[21] J. Joh, A.B. Jenson, W. King, et al., Genomic analysis of the first laboratory-mouse papillomavirus, J. Gen. Virol. 92 (2011) 692–698.
[22] E. Schulz, M. Gottschling, R.G. Ulrich, et al., Isolation of three novel rat and mouse papillomaviruses and their genomic characterization, PLoS One 7 (2012), e47164.
[23] K.D. Quint, R.E. Genders, M.N.C. de Koning, et al., Human Beta-papillomavirus infection and keratinocyte carcinomas, J. Pathol. 235 (2015) 342–354.
[24] C. Borgogna, M.M. Landini, S. Lanfredini, et al., Characterization of skin lesions induced by skin-tropic α- and β-papillomaviruses in a patient with epidermodysplasia verruciformis, Br. J. Dermatol. 171 (2014) 1550–1554.
[25] C. Borgogna, S. Lanfredini, A. Peretti, et al., Improved detection reveals active β-papillomavirus infection in skin lesions from kidney transplant recipients, Modern Pathol. 27 (2014) 1101–1115.
[26] D. Bottalico, Z. Chen, A. Dunne, et al., The oral cavity contains abundant known and novel human papillomaviruses from the Betapapillomavirus and Gammapapillomavirus genera, J. Infect. Dis. 204 (2011) 787–792.
[27] N.M. Cladel, L.R. Budgeon, K.K. Balogh, et al., A novel pre-clinical murine model to study the life cycle and progression of cervical and anal papillomavirus infections, PLoS One 10 (2015), e0120128.
[28] N.M. Cladel, L.R. Budgeon, T.K. Cooper, et al., Secondary infections, expanded tissue tropism, and evidence for malignant potential in immunocompromised mice infected with Mus musculus papillomavirus 1 DNA and virus, J. Virol. 87 (2013) 9391–9395.
[29] J. Hu, L.R. Budgeon, N.M. Cladel, et al., Tracking vaginal, anal and oral infection in a mouse papillomavirus infection model, J. Gen. Virol. 96 (2015) 3554–3565.

[30] A. Handisurya, P.M. Day, C.D. Thompson, et al., Strain-specific properties and T cells regulate the susceptibility to papilloma induction by Mus musculus papillomavirus 1, PLoS Pathog. 10 (2014), e1004314.
[31] J.P. Sundberg, T.M. Stearns, J. Joh, et al., Immune status, strain background, and anatomic site of inoculation affect mouse papillomavirus (MmuPV1) induction of exophytic papillomas or endophytic trichoblastomas, PLoS One 9 (2014), e113582.
[32] M. Gottschling, M. Göker, A. Stamatakis, et al., Quantifying the phylodynamic forces driving papillomavirus evolution, Mol. Biol. Evol. 28 (2011) 2101–2113.
[33] K. van Doorslaer, H.-U. Bernard, Z. Chen, et al., Papillomaviruses: evolution, Linnaean taxonomy and current nomenclature, Trends Microbiol. 19 (2011) 49–50 (author reply 50-1).
[34] S.D. Shah, J. Doorbar, R.A. Goldstein, Analysis of host–parasite incongruence in papillomavirus evolution using importance sampling, Mol. Biol. Evol. 27 (2010) 1301–1314.
[35] T. Brake, P.F. Lambert, Estrogen contributes to the onset, persistence, and malignant progression of cervical cancer in a human papillomavirus-transgenic mouse model, Proc. Natl. Acad. Sci. U. S. A. 102 (2005) 2490–2495.
[36] A.J. Schaeffer, M. Nguyen, A. Liem, et al., E6 and E7 oncoproteins induce distinct patterns of chromosomal aneuploidy in skin tumors from transgenic mice, Cancer Res. 64 (2004) 538–546.
[37] R.R. Riley, S. Duensing, T. Brake, et al., Dissection of human papillomavirus E6 and E7 function in transgenic mouse models of cervical carcinogenesis, Cancer Res. 63 (2003) 4862–4871.
[38] S. Song, A. Liem, J.A. Miller, P.F. Lambert, Human papillomavirus types 16 E6 and E7 contribute differently to carcinogenesis, Virology 267 (2000) 141–150.
[39] I.D. Schaper, G.P. Marcuzzi, S.J. Weissenborn, et al., Development of skin tumors in mice transgenic for early genes of human papillomavirus type 8, Cancer Res. 65 (2005) 1394–1400.
[40] R. Pfefferle, G.P. Marcuzzi, B. Akgül, et al., The human papillomavirus type 8 E2 protein induces skin tumors in transgenic mice, J. Invest. Dermatol. 128 (2008) 2310–2315.
[41] G.P. Marcuzzi, M. Hufbauer, H.U. Kasper, et al., Spontaneous tumour development in human papillomavirus type 8 E6 transgenic mice and rapid induction by UV-light exposure and wounding, J. Gen. Virol. 90 (2009) 2855–2864.
[42] M.S. Campo, Animal models of papillomavirus pathogenesis, Virus Res. 89 (2002) 249–261.

CHAPTER 8

Potential drug targets against HPV and repurposing of existing drugs

Utkarsha Sahu, Prashant Khare, and Ashvini Yadav
Department of Microbiology, All India Institute of Medical Sciences, Bhopal, Madhya Pradesh, India

1. Introduction

Papillomaviruses are a group of small nonenveloped double-stranded DNA (dsDNA) viruses infecting the mucosal tissues of humans and other vertebrate species [1]. Human papillomavirus (HPV) belongs to the family *Papillomaviridae* and possesses an 8 kb genome encoding different early (E) and late (L) proteins. HPV can be broadly divided into five genera—alpha (α), beta (β), gamma (γ), mu (μ), and nu (ν) [2]. The α genus of HPV mainly infects the differentiating keratinocytes of the genital mucosa, whereas the other genus leads to cutaneous level infection [3]. HPVs are further categorized into two categories based on their property to cause malignant progression, namely, high-risk HPV (HR-HPV) and low-risk HPV (LR-HPV). Infection with HR-HPV strains leads to different malignancies such as cervical cancer (CC), anal cancer, and oropharynx and penile cancer [4]. CC associated with HR-HPV (HPV16 and 18) is one of the leading causes of mortality in women worldwide, predominantly in underdeveloped countries [5]. Infection with LR-HPV strains like HPV6 and HPV11 leads to laryngeal papillomatosis or genital warts, a rare but unbearable disease transferred from the mother to the child during birth via viral transmission [6]. The mortality rate associated with HPV-induced malignancies is very alarming. The situation in developing countries is even worse, with approximately 200,000 deaths due to CC every year. This is due to the absence of appropriate screening programs [7].

During the past decade, there has been a significant advancement in our knowledge of papillomavirus infection, pathogenesis, and immune response.

Immunopathology, Diagnosis, and Treatment of HPV-induced Malignancies
https://doi.org/10.1016/B978-0-323-91797-1.00007-7

Primary and secondary prevention programs have a decisive impact on cancer prevention, but they are unavailable for many people at the highest risk. A patient with HPV-induced neoplastic or chronic lesions and malignancies potentially needs therapy. When the surgical removal is not successful or not possible, the development of other approaches is necessitated. This development needs an understanding of potential drug targets and immune responses associated with viral and lesion clearance [8].

The viral proteins serve as potential drug targets to limit viral pathogenesis. The HPV early proteins are vital for viral replication, pathogenesis, and immune evasion. For instance, The E1 and E2 complex is a prerequisite for the initiation of viral replication. Targeting these proteins would aid in limiting the initiation of HPV replication. The HPV oncoproteins E6 and E7 induce malignant transformation of infected cells. A lot of efforts have been made to inhibit these oncoproteins. Significant therapeutic and prophylactic attempts to limit HPV infection were developed by employing the E6 and E7 inhibition [9–13]. In addition to these, E5, the smallest HPV oncoprotein, also plays a crucial role in HPV pathogenesis [14–18]. Recently, different attempts are being carried out to target the E5 protein in order to restrict HPV infection [19,20].

Due to great abrasion rates, high cost, and time constrain of drug development, repurposing of preexisting drugs to treat other diseases is becoming effective since it uses the derisked compounds which have been previously tested to be safe for administration. Repurposing of old drugs will potentially decrease the development cost and time constrain. Several studies have been performed to repurpose different antiviral, antiparasite, antiinflammatory drugs against HPV. Various experimental trials have been proposed for the drug candidate(s) identification and potentially repurposable drugs. In the current book chapter, we have discussed different drug targets and drug repurposing candidates against HPV.

2. Potential drug targets against HPV

2.1 E1 and E2 proteins

The initiation of replication of viral episomal DNA requires proteins E1 and E2. The E2 protein aids the binding of E1 to the ori and induces its helicase activity. There is a formation of a ternary complex of E1-E2-ori that helps in recognition of the origin of replication and recruitment of other E1 molecules capable of unwinding the viral DNA [21–23]. To date, several attempts have been carried out to prevent the formation of this complex. However, only a few of them proved to be effective in vivo [24]. Indandiones, a group of small molecules, explicitly restricted the E1-E2 interaction and therefore also restricted viral DNA replication [25,26]. The inhibitors of this class were recognized by high throughput screening of different compounds for their capability to limit the E1-E2 dimer [26]. These inhibitors had an indandione system substituted tetrahydrofuran ring and bind with E2 TAD domain. This TAD domain is responsible for binding with E1. The crystal structure of HPV11 TAD and its interaction with inhibitor indandione revealed that how the binding and disruption of the E1-E2 complex is done by indandione inhibitors [27].

The unwinding and ATPase activities are found in the C-terminus of E1 that also binds with E2 [28]. It has been reported that the binding of E2 with E1 interferes with the E1 ATPase activity. Also, ATP binding weakens the E1-E2 interaction and deteriorates their binding to the origin of replication [28]. These findings proposed that apart from aiding the E1 helicase activity, ATP acts like a molecular trigger to promote hexamerization of E1 and breaks the E1-E2 dimer [28]. By itself, aiming the ATPase activity and/or ATP binding activity can be a practical step to inhibit DNA replication (Fig. 1).

Small molecules inhibiting the HPV6 E1 ATPase activity were identified via high-throughput screening [29]. Biphenylsulfonacetic acid analogs containing lead molecules are categorized by the sulfonyl acetic acid [29]. These molecules exhibited inhibition of ATP binding activity by an allosteric mechanism [30]. Therefore, the inhibitor activity is dependent on the concentration of ATP.

2.2 E5 protein

The HPV E5 protein is the smallest oncoprotein that is vital for HPV persistence and transformation of infected cells. E5 activates several cellular pathways during an early stage of CC [15]. E5 works along with other major oncoproteins, E6 and E7, in promoting hyperproliferation and malignant progression of HPV-infected cells. In addition, it binds with the B cell receptor-associated protein 31 in ER and panels trafficking of proteins. It also downregulates the MHC I expression and helps in evading the host immunity [16,17,31–33]. Different studies have emphasized the importance of targeting the E5 oncoprotein. In vivo studies showed that HPV16 E5 delivered by the adenovirus vector reduced the tumor growth, whereas E5 vaccine triggered immunity against tumors by the $CD8^+$ cytotoxic T (CTL) cells [34]. The study demonstrated that an intramuscular injection of the HPV16 E5 gene carrying adenovirus into the C57BL/6 (H-2^b) mice reduced the growth of E5 expressing tumors. Also, this protection occurred via CTL but not Th cells in the in vitro assays. The study on knockout (KO) mice with T cell deficiency suggested that CTL-dependent tumor suppression is dependent on $CD8^+$ T cells [34]. Thus, it can be proposed that E5 can be an antigen for tumor rejection. The same group further went to prove that E5 was an immune-stimulating peptide and vaccination using this peptide significantly reduced tumor growth [35].

The expression of E5 is mainly upregulated during the early stages of viral infection and premalignant lesions [36]. Thus, targeting E5 could avoid the advancement of infection and conversion of lesions to invasive cervical cancer. ET1, EGF-R, or COX-2 inhibitors might be useful to regulate E5 activity in the premalignant lesions. At last, E5 can be considered like the SV40VP4 viroporin [37] and, thus, vulnerable to compounds affecting similar structures of other viruses.

2.3 E6 protein

The E6 oncoprotein of HPV is crucial for viral persistence, evasion of host immune response, and malignant progression. E6 interacts with p53 and recruits different ubiquitinating enzymes for its degradation, thereby preventing apoptosis of infected cells [38,39]. Several small molecules, natural compounds, and drugs have been tested to inhibit HPV infection by targeting E6 proteins (Fig. 1).

In vivo and in vitro analysis showed that a combined treatment of dihydrotanshinone I (DI) and irradiation (IR) increased apoptosis of CC cells by downregulating the E6 oncoprotein and caspases [40]. DI positively regulated the p21 expression and downregulated the p34 and cyclin B1 expression in HeLa cells exposed to IR. This association was associated with cell cycle arrest during the G2 phase. In addition, an upsurge in the activity of caspase-3 was also reported. Later, these HeLa cells were used as xenografts in the nude mice followed by IR and DI treatment, and a significant reduction in tumor progression was observed without organ or general toxicity [40]. Polyphenolic compounds have been widely reported for their antiviral properties [41,42]. Polyphenolic compounds, pterostilbene and resveratrol, possess anticancer properties in vivo and in vitro and antitumor properties. In vitro experiments on murine TC1 cells positive for HPV E6 and E7 proteins. In vivo results suggested that TC1 tumors significantly downregulated after resveratrol or pterostilbene treatment. In addition, these compounds downregulated the E6 oncoprotein. The reduction in tumor size

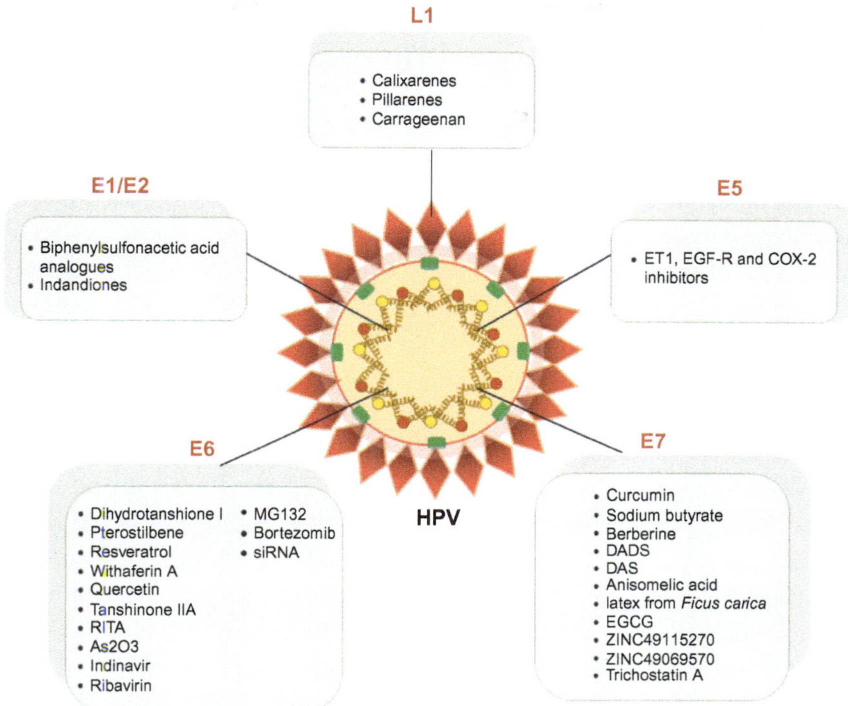

Fig. 1 Different drug targets of HPV and potential inhibitors.

after pterostilbene treatment was accompanied by apoptosis of tumor cells, upregulation of caspase-3, whereas after resveratrol treatment, major events were cell cycle arrest and downregulation of PCNA. Thus, these compounds can tackle the HPV infection by targeting viral oncoproteins [43].

Withaferin A (WA) is a component derived from the medicinal plant *Withania somnifera* that has been reported to inhibit different malignancies such as prostate, pancreatic, and breast cancer [44–46]. During in vitro and in vivo analysis of WA in CC, WA reduced the proliferation of CaSki cell line downregulated E6 and E7, accumulated P53, cell cycle arrest at the G2-M phase, increased p21cip/waf, decreased SAT3 levels, and its phosphorylation and also altered the expression of p53 apoptotic markers—caspase-3, Bax, and BCL2 [47]. In vivo, WA leads to reduction of about 70% tumor size in athymic nude mice [47].

Quercetin, a flavonoid polyphenolic compound, is used against different malignancies such as hepatocellular carcinoma where quercetin restricted cell proliferation, induced cell cycle arrest, DNA fragmentation, apoptosis, upregulated p53 expression, and phosphorylation in SiHa cells and HeLa cells [48]. However, quercetin had no effect on the HPV E6 expression in CC cells. Thus, the p53 upregulation was independent of the E6 expression [49].

Tanshinone IIA (Tan IIA), a diterpenoid naphthoquinone from *Salvia* species, repressed human CC proliferation in C33a, SiHa, CaSki, and HeLa cell lines [50]. In HPV-infected CaSki cells, TanIIA downregulated the E6 and E7 expression, led to cell cycle arrest at the S phase, induced p53 accumulation and modified the expression of p53 target genes, and modulated retinoblastoma protein (pRB) and related proteins [50]. 3′-O-methylnordihydroguaiaretic acid, a lignin from the creosote bush, can inhibit the E6 expression by targeting its promoter [10]. The repressed E6 expression resulted in a stable and transcriptionally active p53 protein [10].

A small-molecule reactivation of p53 and induction of tumor cell apoptosis (RITA) inhibit HPV 16/18 by inducing p53 accumulation and retain the tumor suppression function in HPV 16/18 infected cells by targeting E6 associated with proteasomal degradation [51]. RITA limits ubiquitination of p53 by preventing its associations with the E6-associated protein, vital for HPV E6-mediated degradation. RITA triggered proapoptotic p53 transcription and repressed the expression of different proproliferative factors such as PUMA, BAX, and Noxa and repressed CDC2, CDC25C, and CyclinB1, ensuring cell cycle arrest and p53-dependent apoptosis. In vivo, RITA displayed CC xenografts suppression [51]. Another approach to target E6 is treatment with arsenic trioxide (As_2O_3) that eventually leads to HPV E6 downregulation and p53 upregulation and triggered cell cycle arrest in S-G2/M stage and apoptosis [52].

The E6 oncoprotein regulates translation of eukaryotic translation initiation factor 4E (eIF4E) by targeting the downstream targets—MEK/ERK pathway and PI3K/AKT/mTORC1 pathway [53]. The upregulated eIF4E expression is vital for HPV-associated cancer. Indinavir and ribavirin are eIF4E inhibitors [54]. In vitro analysis of these compounds on HeLa cells decreased their viability, decreased MMP secretion, and induced apoptosis. In vivo analysis of indinavir and ribavirin immunocompromised C57BL/6 mice with human xenograft reduced tumor volume [54]. An additional study by Oei et al. introduced a hyperthermia therapy for improving outcomes. The mechanism of hyperthermia therapy includes hyperthermia at 42°C for 60 min that abolishes p53 and E6 interactions resulting in accretion of p53 and aiding p53-mediated apoptosis and G2 cell cycle arrest. These results were observed in in vitro conditions in different cell lines, in xenografts, and in biopsies from CC patients [55].

The E6 protein marks p53 for proteasome-mediated degradation. Thus, using proteasome inhibitors to prevent the proteasome-ubiquitin complex is also tested in different cell lines. The MG132 proteasome inhibitor upregulated transcriptional activity and p53 protein levels in human cell lines [56]. Bortezomib has an affinity for chymotrypsin-like proteasome activity [57] and sensitizes SiHa cells for radiotherapy by NF-kB inhibition in tumor xenografts and induced caspase 3 cleavage. Also, it limited the tumor growth by decreasing vascular endothelial growth factor (VEGF) production [58,59]. Recent studies have presented that an amalgamation of radiotherapy and bortezomib was well tolerated by advanced cancer patients [60].

RNAi involves inhibition of particular mRNA by homologous double-stranded RNA (dsRNA) that significantly inhibits the protein expression of mRNA [61,62]. It is reported that small interfering dsRNA (siRNA) of about 21 bases reduced the E6 and E7 expression, and inhibition of these oncoproteins triggered HeLa cell senescence [63].

2.4 E7 protein

The E7 oncoprotein of HPV is generally expressed in the lower layers of host epithelium and restricts the cells in the S phase of the cell cycle providing suitable environment for viral replication and proliferation [64]. E7 coordinates with the Rb protein and disturbs the binding of Rb and transcription factor E2F, releasing E2F and transcription of the S phase promoting factors. This combined association of E6 and E7 oncoproteins overtakes the cell cycle checkpoints and aids HPV replication [65,66]. Along with E6, E7 is also a crucial drug target in order to limit the HPV infection. Different compounds and RNAi approaches were carried out to target the E7/E6 axis in HPV-induced malignancies (Fig. 1).

Curcumin is a polyphenolic compound derived from turmeric rhizomes, *Curcuma longa*, and is found to possess anticancer effects [67]. Curcumin acts via different cellular pathways and prompts different effects such as antimutagenic effects, antiinflammatory effects, and antiangiogenesis effects [68–70]. Crucially, it has no toxic effects on healthy organs even at high doses [71]. After 24 h of curcumin treatment, the expression of E7 along with E6 was significantly downregulated. In addition, it increased the expression of Rb, p53, and protein tyrosine phosphatase nonreceptor type 13 (PTPN13) [71]. Another study by Divya et al. also reported the inhibition of E7 along with E6 by curcumin [72]. Berberine (5, 6-dihydro-9, 10-dimethoxybenzo[g]-1, 3-benzodioxole5,6-aquinolizum), an alkaloid isolated from medicinal plant *Berberis* (*B. aristata, B. vulgaris, B. aquifolium,* etc.), was reported to possess anticancer and anti-inflammatory properties [73–75]. In addition, it also possesses anti-HPV properties by targeting the E7 oncoprotein [76].

Organosulfur compounds, diallyl disulfide (DADS) and diallyl sulfide (DAS), from garlic (*Allium sativum*) exhibited antiproliferative results on CaSki cells [12]. DADS and DAS restricted the expression of viral oncoproteins, E7 and E6, and restored the p53 function. In addition, they lead to cell cycle arrest at the G0-G1 phase by downregulating CDK4 and cyclin D1 and upregulating the p27KIP1 and p21WAF1/CIP1 CDK inhibitors [12].

Anisomelic acid, a diterpenoid, restricted E7 and E6 oncoproteins in the CC cell line [77]. As a result, p53 and its responsive gene p21 were increased that caused cell cycle arrest in the G2-M phase. Anisomelic acid also induced cell apoptosis by the depleting cellular inhibitor of apoptosis protein 2 (cIAP2). In vivo analysis of anisomelic acid in the chick embryo chorioallantoic membrane prevented tumor growth in SiHa cells [77].

The antimycin analogs derived from different species of marine *Streptomyces* inhibitor HPV E7 and E6 induces ROS-mediated ubiquitin and proteasome system in HeLa cells [78]. Fruit latex from *Ficus carica* downregulated the E7 and E6 expression in CaSki cell lines positive with HPV. In addition, it also reduced HPV-associated warts, indicating its potential use to cure HPV lesions [79].

(−)-Epigallocatechin-3-gallate (EGCG), a green tea molecule, and its analogs are also reported to inhibit E7 and E6 molecules [64,80]. EGCG restricts the expression of aromatase and estrogen receptor α (ERα) that indirectly limits the E7 and E6 expression and growth of cervical cells [80]. Compound ZINC49115270 and ZINC49069570 analogs of EGCG bind with E7 and inhibit its activity [64]. The immortalizing and hyperproliferative properties of E7 are also contributed due to its capability to target histone deacetylases class I (HDAC1 and HDAC2) [81]. Therefore, the HDAC inhibitors can be used to restrict HPV pathogenesis.

These inhibitors were initially reported to induce apoptosis and growth arrest in different cancer cells such as ovarian, prostate, and colon [82]. Two HDAC inhibitors, sodium butyrate and trichostatin A, are reported to limit HPV immortalized cells in different ways. Firstly, they restore the inhibitory activity of Cdk2 by activating CDK1 inhibitors, p27 and p21 [83,84]. In addition, they also restore the function of IRF-1 in the host immune response [138]. Recently, HDAC inhibitor sodium butyrate was reported to prompt apoptosis and by targeting the HPV E7-Rb-E2F pathway that was independent of p53 [85].

The RNAi studies on HeLa cells reported that E7 siRNA considerably reduced the replication of cellular DNA. There was a remarkable decrease in DNA synthesis that continued for 12 days after transfection. There was no downregulation in DNA synthesis in HPV^{-} C33A cell line, suggesting that E7 siRNA repressed cell growth by interfering with the gene expression [63]. Another study on siRNA-mediated E7 silencing triggered Rb hypophosphorylation. The presence of E7-shRNA upregulated p53 and p21 protein. Additionally, the hyperphosphorylated Rb was reduced during the presence of E7 shRNA that suppressed the HPV16 E7 protein expression [86].

2.5 L1 protein

The L1 protein is the major capsid protein crucial for viral attachment and entry and is capable of forming virus-like particles on its own. The L1 protein is highly immunogenic and is expressed in later stages of the viral life cycle to avoid host immune surveillance. Although the L1 protein is widely targeted in the different attempts of vaccine development, the role of L1 as a potential target is yet to be explored. There is currently no strong experimental evidence of drugs present that target the HPV L1 protein. However, a study performed on 14 peptides revealed its ability to restrict L1 pentamer formation in HPV16. Similar inhibitory effects were shown for the L1 protein of HPV58 [87].

Calixarenes and pillararenes are types of synthetic macrocycles that inhibit L1 pentamer formation by binding to the lysine and arginine residues. However, as compared to calixarenes, pillararenes were better inhibitors in terms of binding affinity [88]. The carrageenan extract (ι-Carrageenan) from Irish moss specifically binds with the LI protein of HPV and prevents its attachment to host epithelial cells. This is due to its resemblance with the heparin protein required by HPV to enter host cells. In addition, ι-carrageenan also restricted HPV infection after the attachment of virions. ι-Carrageenan was also effective in different HPV types, including 6, 16, 18, 31, and 45; however, it had no activity for HPV5 [1]. These findings paved the way for developing a new category of HPV inhibitors that target the viral capsid proteins.

3. Repurposing of existing drugs against HPV

Repurposing of existing drugs for treating rare and common diseases has recently gained the huge attention of different researchers across the globe. The significant factors for this are no-risk, cost-effectiveness, and speedy delivery due to preestablished clinical trials. The status of HPV therapeutic research is also focusing on repurposing the existing drugs against different viral, bacterial, and parasitic infections (Table 1). Also, the class of antiinflammatory drugs is becoming popular to target virus-induced inflammation [101].

3.1 Antiviral drugs

The HPV E6 protein prompts the eIF4E factor expression, and thus, restricting eIF4E can inhibit malignant progression of HPV [102]. Two antiviral agents, indinavir and ribavirin, are recognized to inhibit the eIF4E activity [90]. Their anti-HPV activity was tested in HeLa cells. Ribavirin and indinavir were cytotoxic to the HeLa cells and induced apoptosis. They also downregulated the matrix metallopeptidase (MMP)-2 and MMP-9 expression that are crucial for metastasis. In addition, the in vivo study on immunocompromised mice revealed that both of the molecules limited the tumor progression [54]. Cidofovir or 1-[(*S*)-3-hydroxy-2-(phosphonylmethoxy)-propyl] cytosine is a nucleoside phosphonate with antiviral activity against HIV patients infected with cytomegalovirus (CMV) retinitis [91]. Cidofovir is also reported to downregulate the HPV E6 and E7 oncoprotein expression and metastasis in HPV^+ tumors [11]. In addition, a study performed on CIN3 affected women alternate day treatment of topical cidofovir completely regressed CIN3 in 47% of women after three treatments [92].

Recently, four different anti-HIV drugs, ritonavir, lopinavir, saquinavir, and nelfinavir, were tested for their anti-HPV activities. They selectively targeted the HPV E6 and E7 oncoproteins in HPV^+ cell lines and in the $aHPV16^+$ 3D model of NIKS16 cells. Depletion of E6 and E7 displayed an increased p53 expression and apoptosis in CaSki cells [93]. Another study by Hampson et al. also reported similar findings supporting the use of anti-HIV-inhibitors to target HPV [94].

3.2 Antibiotics

The antibiotic doxycycline is a broad-spectrum antibiotic that is monitored during bacterial infections. It works by inhibiting bacterial protein synthesis [103]. Recently, it was also analyzed in HPV^+ and HPV^- cervical cancer cells both in vivo and in vitro [89]. Doxycycline inhibited the proliferation and induced apoptosis in CC cell lines in a dose- and time-dependent manner via a caspase-dependent mechanism. One study showed the mode of action of doxycycline revealed that it targets the rate of oxygen consumption and glycolysis and significantly reduces ATP in CC cells. In xenograft

Table 1 Different classes of repurposed drugs to target HPV infection and associated malignancies.

Category	Drugs/compounds	Original target	Mode of action	References
Antibiotic	Doxycycline	Bacteria	• Inhibits proliferation of infected cells and induces apoptosis by caspase-mediated pathway • Targets the oxygen consumption of cell by reducing ATP in CC cells • Prevents tumor growth in vivo	[89]
Antivirals	Indinavir Ribavirin	HIV Hepatitis C virus Respiratory syncytial virus	• Cytotoxic activity in HeLa cells • Induces apoptosis • Targets eIF4E transcription factor and limits proliferation • Downregulates the MMP expression essential for metastasis • Prevents in vivo tumor growth	[54,90]
	Cidofovir	HIV, CMV	• Downregulation of metastasis and HPV E6 and E7 oncoproteins • Regression of CIN3	[91,92]
	Ritonavir Lopinavir Saquinavir Nelfinavir	HIV	• Downregulation of E6 and E7 oncoproteins in HPV^+ cell lines and in the $HPV16^+$ 3D model of NIKS16 cells • Increases the p53 expression and apoptosis in CaSki cells	[93,94]
Antiinflammatory drugs	Meclofenamic acid	COX pathway in different malignancies	• Possesses antiinflammatory property and induces apoptosis in cancer cells	[95]
	Celecoxib		• Induces apoptosis in different CC cell lines	[96]
	Aspirin Acetaminophen		• Limits the COX-mediated inflammation	[96,97]
Antiparasitic drugs	Artemisinin	Malaria parasite	• Inhibits the E6 and E7 oncoproteins of HPV39 • Inhibits cancer cell proliferation by limiting the telomerase subunits and inducing apoptosis	[98,99]
	Artesunate		• Anti-HPV property targeting the intraepithelial neoplasia	[100]

of HeLa cells, it limited the tumor growth in severe combined immune-deficient (SCID) mice. Altogether, these results suggest that doxycycline can be repurposed for treating HPV-induced CC [89].

3.3 Antiparasitic drugs

Antimalarial drug artemisinin was reported to inhibit the E6 and E7 oncoproteins of HPV39. It inhibited cancer cell proliferation by limiting the telomerase subunits and inducing apoptosis [98]. Furthermore, the p53 expression upregulated, and the inhibitor of p53—Mdm2—was down-regulated, suggesting that artemisinin employs p53-dependent apoptosis in cancer cells [98]. Another study by Disbrow et al. also supported the anti-HPV role of the chemical derivative of artemisinin in in vitro and in vivo conditions [99].

Artesunate, another antimalaria drug, is currently under investigation in the United States for its anti-HPV property targeting the intraepithelial neoplasia [100].

3.4 Antiinflammatory drugs

Inflammation is the major event during HPV infection. Dysregulated inflammatory response during HPV infection contributes to skin lesions and malignant progression [104]. The inflammation is triggered mainly by the cyclooxygenase 2 (COX-2) enzyme, and it has been targeted to counter different malignancies [105,106]. Nonsteroidal antiinflammatory drugs (NSAIDs) are a class of drugs that targets the COX activity and were tested in HPV+ SiHa and INBL CC cell lines and mouse TC-1 cell line [95]. Among different tested drugs, meclofenamic acid showed highest cytotoxicity, whereas sulindac, nimesulide, and celecoxib had partial cytotoxicity [95]. Celecoxib, another COX inhibitor, possesses antiinflammatory properties and induces apoptosis in cancer cells. Celecoxib also induced apoptosis in different CC cell lines—C33A, CaSki, and HeLa cell lines. The microarray cDNA chip screening of 225 genes highlighted that growth arrest and DNA damage-inducible gene (GADD153) was highly differentiated after celecoxib treatment in all the CC cell lines. GADD153 is a transcription factor that is mainly associated with cell apoptosis. These findings suggested that celecoxib limited CC via GADD153 [96].

Aspirin and acetaminophen, other antiinflammatory drugs, were evaluated to inhibit the risk of HPV-induced CC. Both aspirin and acetaminophen limited the COX-mediated inflammation and induced apoptosis [97].

4. Conclusion

Infection with HR-HPV strains remains the principal cause of CC and other malignancies in women across the world. HPV-induced CC is the fourth leading cancer in women, causing severe mortality worldwide. The current vaccines against HPV are mainly prophylactic, i.e., they can prevent the HPV infection. However, they are insufficient in clearing the preestablished HPV infection, associated lesions, and malignant progression. The current treatment options against HPV-induced malignancies include radiation, surgery, and chemotherapy. These therapies affect both infected and noninfected cells. Thus, the present time calls for better and effective molecular-targeted strategies. This includes the identification of potential drug targets for treating HPV infection.

The initiation of viral replication requires the binding of the E1/E2 complex at the origin of replication in the viral genome. The helicase activity of E1 is crucial for the unwinding of viral DNA during replication. The E1-E2 complex is targeted to limit the viral genome replication. Compounds like biphenylsulfonacetic acid analogs containing lead molecules and indandiones [25,26] are reported to target the ATPase activity of this complex and thereby limiting the HPV replication. The E5 oncoprotein of HPV is required to evade host immune response and aids E6/E7-mediated transformation of HPV-infected cells [16,17,31–33]. However, the study of compounds and small-molecule inhibitors against E5 is limited [37]. Further research on drugs targeting the E5 protein would be beneficial against HPV infection. The oncogenic role of E6-E7 viral proteins via modulation of host immune response and cell cycle regulation is well established [65,66,107]. E6 binds with the p53 tumor suppressor protein and mediates the recruitment of different ubiquitination enzymes for p53 degradation to prevent cell death. The E7 protein disturbs the binding of Rb and transcription factor E2F, releasing E2F and transcription of S phase-promoting factors. This combined association of E6 and E7 oncoproteins supersedes the cell cycle checkpoints and aids HPV replication [65,66]. E6 and E7 are predominantly the major targets to develop therapeutic and prophylactic measures against HPV. A lot of studies were carried to target these oncoproteins. Different compounds like dihydrotanshinone I, pterostilbene and resveratrol, withaferin A, quercetin, curcumin, DADS, DAD [12,40,44–46,48,68–70,72], etc., are tested against E6-E7 proteins. In addition, different RNAi approaches were developed to aim the E6/E7-mediated oncogenic progression of HPV-infected cells [63,86]. Also, different in silico therapeutic vaccine candidates were constructed taking these oncoproteins in account [9,108]. The capsid proteins of HPV (L1 and L2) are necessary for viral attachment and entry. The interaction of HPV L1 and L2 proteins with the extracellular matrix components is responsible for the intake of the virus into host cells [109–111]. Targeting these capsid proteins would help to prevent the first step of viral pathogenesis, i.e., attachment and entry. Although the L1 capsid protein is highly immunogenic till date, there are very less drugs against HPV L1 as compared to the oncoproteins E6 and E7 although their use in in silico vaccine design is being carried out [112,113]. There is a need to develop more effective therapeutics against viral attachment and entry. The management of HPV-induced cancer is still a task to mankind that requires the discovery and development of new therapeutics [89]. The identification, discovery, and development of potential druggable targets and novel drug compounds is a laborious process. Repurposing of preexisting drugs may surpass this procedure and aid the rapid translation of hypothesis-motivated science into the clinical outcomes [114,115]. Different types of antiviral, antiparasitic, and antiinflammatory drugs were tested for limiting HPV infection and related malignancies [90,95–99,116].

Indinavir and ribavirin are another set of antiviral drugs that exhibited anti-HPV activity in HeLa cells [90]. Antimalaria drug artemisinin inhibited the E6 and E7 oncoproteins of HPV39 by limiting the telomerase subunits and inducing apoptosis [98]. Furthermore, the p53 expression upregulated and the inhibitor of p53, i.e., Mdm2, was downregulated, suggesting that artemisinin employs p53-dependent apoptosis in cancer cells [98]. Another study also supported the anti-HPV role of the chemical derivative of artemisinin in in vivo and in vitro conditions [99]. Artesunate, another antimalaria drug, is currently under investigation in the United States for targeting HPV-induced intraepithelial neoplasia [100].

Inflammation is the crucial event for the establishment and perseverance of HPV infection that finally leads to CIN and malignant progression of infected cells [116,117]. Different antiinflammatory drugs such as meclofenamic acid, sulindac, nimesulide [95], celecoxib [96], aspirin, and acetaminophen [97] were used to target this viral-induced COX-mediated inflammatory response.

Currently, there is a requirement for better access to industry-sponsored clinical trial data. This would facilitate us to explore new findings to pave new opportunities for better drug development [115]. Also, better funding opportunities, technology, and sharing of repurposing libraries are required.

The cumulative efforts of different branches of science across the world would help to develop better therapeutics against HPV-induced malignancies. Comprehensive knowledge of varying drug targets and repurposing strategies would lead to the development of new therapeutics against LR and HR-HPV strain-associated ailments.

References

[1] C.B. Buck, et al., Carrageenan is a potent inhibitor of papillomavirus infection, PLoS Pathog. 2 (7) (2006), e69.
[2] Y.J. Choi, J.S. Park, Clinical significance of human papillomavirus genotyping, J. Gynecol. Oncol. 27 (2) (2016), e21.
[3] V. Bouvard, et al., A review of human carcinogens—part B: biological agents, Lancet Oncol. 10 (4) (2009) 321–322.
[4] A. Bansal, M.P. Singh, B. Rai, Human papillomavirus-associated cancers: a growing global problem, Int. J. Appl. Basic Med. Res. 6 (2) (2016) 84–89.
[5] D.M. Parkin, et al., Global cancer statistics, 2002, CA Cancer J. Clin. 55 (2) (2005) 74–108.
[6] L. Gissmann, et al., Human papillomavirus types 6 and 11 DNA sequences in genital and laryngeal papillomas and in some cervical cancers, Proc. Natl. Acad. Sci. U. S. A. 80 (2) (1983) 560–563.
[7] A. Batson, F. Meheus, S. Brooke, Chapter 26: innovative financing mechanisms to accelerate the introduction of HPV vaccines in developing countries, Vaccine 24 (Suppl 3) (2006) S3/219–25.
[8] P.L. Stern, et al., Therapy of human papillomavirus-related disease, Vaccine 30 Suppl 5 (5) (2012) F71–F82.
[9] B. Jabbar, et al., Antigenic peptide prediction from E6 and E7 oncoproteins of HPV types 16 and 18 for therapeutic vaccine design using immunoinformatics and MD simulation analysis, Front. Immunol. 9 (3000) (2018).
[10] K.L. Allen, et al., A plant lignan, 3'-O-methyl-nordihydroguaiaretic acid, suppresses papillomavirus E6 protein function, stabilizes p53 protein, and induces apoptosis in cervical tumor cells, Mol. Carcinog. 46 (7) (2007) 564–575.
[11] A. Amine, et al., Novel anti-metastatic action of cidofovir mediated by inhibition of E6/E7, CXCR4 and rho/ROCK signaling in HPV tumor cells, PLoS One 4 (3) (2009), e5018.
[12] I.A. Ansari, et al., Organosulphur compounds induce apoptosis and cell cycle arrest in cervical cancer cells via downregulation of HPV E6 and E7 oncogenes, Anticancer Agents Med Chem. 21 (3) (2021) 393–405.
[13] J.D. Baleja, et al., Identification of inhibitors to papillomavirus type 16 E6 protein based on three-dimensional structures of interacting proteins, Antiviral Res. 72 (1) (2006) 49–59.
[14] M. Gruener, et al., The E5 protein of the human papillomavirus type 16 down-regulates HLA-I surface expression in calnexin-expressing but not in calnexin-deficient cells, Virol. J. 4 (1) (2007) 116.
[15] M.K. Kim, et al., Human papillomavirus type 16 E5 oncoprotein as a new target for cervical cancer treatment, Biochem. Pharmacol. 80 (12) (2010) 1930–1935.
[16] J.A. Regan, L.A. Laimins, Bap31 is a novel target of the human papillomavirus E5 protein, J. Virol. 82 (20) (2008) 10042–10051.
[17] S.W. Straight, B. Herman, D.J. McCance, The E5 oncoprotein of human papillomavirus type 16 inhibits the acidification of endosomes in human keratinocytes, J. Virol. 69 (5) (1995) 3185–3192.

[18] L.F. Wetherill, et al., High-risk human papillomavirus E5 oncoprotein displays channel-forming activity sensitive to small-molecule inhibitors, J. Virol. 86 (9) (2012) 5341–5351.
[19] A. Namvar, et al., Development of HPV(16,18,31,45) E5 and E7 peptides-based vaccines predicted by immunoinformatics tools, Biotechnol. Lett. 42 (3) (2020) 403–418.
[20] H.A. Panahi, et al., A comprehensive in silico analysis for identification of therapeutic epitopes in HPV16, 18, 31 and 45 oncoproteins, PLoS One 13 (10) (2018) e0205933.
[21] E.T. Fouts, et al., Biochemical and electron microscopic image analysis of the hexameric E1 helicase, J. Biol. Chem. 274 (7) (1999) 4447–4458.
[22] S. Titolo, et al., Identification of domains of the human papillomavirus type 11 E1 helicase involved in oligomerization and binding to the viral origin, J. Virol. 74 (16) (2000) 7349–7361.
[23] J. Sedman, A. Stenlund, The papillomavirus E1 protein forms a DNA-dependent hexameric complex with ATPase and DNA helicase activities, J. Virol. 72 (8) (1998) 6893–6897.
[24] C.M. D'Abramo, J. Archambault, Small molecule inhibitors of human papillomavirus protein - protein interactions, Open Virol J 5 (2011) 80–95.
[25] C. Yoakim, et al., Discovery of the first series of inhibitors of human papillomavirus type 11: inhibition of the assembly of the E1-E2-origin DNA complex, Bioorg. Med. Chem. Lett. 13 (15) (2003) 2539–2541.
[26] P.W. White, et al., Inhibition of human papillomavirus DNA replication by small molecule antagonists of the E1-E2 protein interaction, J. Biol. Chem. 278 (29) (2003) 26765–26772.
[27] Y. Wang, et al., Crystal structure of the E2 transactivation domain of human papillomavirus type 11 bound to a protein interaction inhibitor, J. Biol. Chem. 279 (8) (2004) 6976–6985.
[28] P.W. White, et al., Characterization of recombinant HPV6 and 11 E1 helicases: effect of ATP on the interaction of E1 with E2 and mapping of a minimal helicase domain, J. Biol. Chem. 276 (25) (2001) 22426–22438.
[29] A.M. Faucher, et al., Discovery of small-molecule inhibitors of the ATPase activity of human papillomavirus E1 helicase, J. Med. Chem. 47 (1) (2004) 18–21.
[30] P.W. White, et al., Biphenylsulfonacetic acid inhibitors of the human papillomavirus type 6 E1 helicase inhibit ATP hydrolysis by an allosteric mechanism involving tyrosine 486, Antimicrob. Agents Chemother. 49 (12) (2005) 4834–4842.
[31] D. DiMaio, D. Mattoon, Mechanisms of cell transformation by papillomavirus E5 proteins, Oncogene 20 (54) (2001) 7866–7873.
[32] P. Leechanachai, et al., The E5 gene from human papillomavirus type 16 is an oncogene which enhances growth factor-mediated signal transduction to the nucleus, Oncogene 7 (1) (1992) 19–25.
[33] M.S. Cortese, G.H. Ashrafi, M.S. Campo, All 4 di-leucine motifs in the first hydrophobic domain of the E5 oncoprotein of human papillomavirus type 16 are essential for surface MHC class I downregulation activity and E5 endomembrane localization, Int. J. Cancer 126 (7) (2010) 1675–1682.
[34] D.W. Liu, et al., Induction of CD8 T cells by vaccination with recombinant adenovirus expressing human papillomavirus type 16 E5 gene reduces tumor growth, J. Virol. 74 (19) (2000) 9083–9089.
[35] Y.-F. Chen, et al., Cytotoxic-T-lymphocyte human papillomavirus type 16 E5 peptide with CpG-oligodeoxynucleotide can eliminate tumor growth in C57BL/6 mice, J. Virol. 78 (3) (2004) 1333–1343.
[36] L. Lorenzon, et al., In vivo HPV 16 E5 mRNA: expression pattern in patients with squamous intra-epithelial lesions of the cervix, J. Clin. Virol. 52 (2) (2011) 79–83.

[37] S. Raghava, et al., The SV40 late protein VP4 is a viroporin that forms pores to disrupt membranes for viral release, PLoS Pathog. 7 (6) (2011), e1002116.
[38] T.M. Wise-Draper, S.I. Wells, Papillomavirus E6 and E7 proteins and their cellular targets, Front. Biosci. 13 (2008) 1003–1017.
[39] A. Duenas-Gonzalez, et al., New molecular targets against cervical cancer, Int. J. Womens Health 6 (2014) 1023–1031.
[40] Y. Ye, et al., Combination treatment with dihydrotanshinone I and irradiation enhances apoptotic effects in human cervical cancer by HPV E6 down-regulation and caspases activation, Mol. Cell. Biochem. 363 (1–2) (2012) 191–202.
[41] P. Khare, et al., Current approaches for target-specific drug discovery using natural compounds against SARS-CoV-2 infection, Virus Res. 290 (2020) 198169.
[42] M. Musarra-Pizzo, et al., The antimicrobial and antiviral activity of polyphenols from almond *(Prunus dulcis* L.) Skin, Nutrients 11 (10) (2019) 2355.
[43] K. Chatterjee, et al., Dietary polyphenols, resveratrol and pterostilbene exhibit antitumor activity on an HPV E6-positive cervical cancer model: an in vitro and in vivo analysis, Front. Oncol. 9 (352) (2019).
[44] S. Srinivasan, et al., Par-4-dependent apoptosis by the dietary compound withaferin a in prostate cancer cells, Cancer Res. 67 (1) (2007) 246–253.
[45] Y. Yu, et al., Withaferin a targets heat shock protein 90 in pancreatic cancer cells, Biochem. Pharmacol. 79 (4) (2010) 542–551.
[46] S.D. Stan, et al., Withaferin a causes FOXO3a- and Bim-dependent apoptosis and inhibits growth of human breast cancer cells in vivo, Cancer Res. 68 (18) (2008) 7661–7669.
[47] R. Munagala, et al., Withaferin a induces p53-dependent apoptosis by repression of HPV oncogenes and upregulation of tumor suppressor proteins in human cervical cancer cells, Carcinogenesis 32 (11) (2011) 1697–1705.
[48] S. Tanigawa, M. Fujii, D.X. Hou, Stabilization of p53 is involved in quercetin-induced cell cycle arrest and apoptosis in HepG2 cells, Biosci. Biotechnol. Biochem. 72 (3) (2008) 797–804.
[49] A.F. Clemente-Soto, et al., Quercetin induces G2 phase arrest and apoptosis with the activation of p53 in an E6 expression-independent manner in HPV-positive human cervical cancer-derived cells, Mol. Med. Rep. 19 (3) (2019) 2097–2106.
[50] R. Munagala, et al., Tanshinone IIA inhibits viral oncogene expression leading to apoptosis and inhibition of cervical cancer, Cancer Lett. 356 (2 Pt B) (2015) 536–546.
[51] C.Y. Zhao, et al., Rescue of p53 function by small-molecule RITA in cervical carcinoma by blocking E6-mediated degradation, Cancer Res. 70 (8) (2010) 3372–3381.
[52] X. Wen, et al., Arsenic trioxide induces cervical cancer apoptosis, but specifically targets human papillomavirus-infected cell populations, Anticancer Drugs 23 (3) (2012) 280–287.
[53] V. Morales-Garcia, et al., The high-risk HPV E6 proteins modify the activity of the eIF4E protein via the MEK/ERK and AKT/PKB pathways, FEBS Open Bio 10 (12) (2020) 2541–2552.
[54] S. Sharma, R. Baksi, M. Agarwal, Repositioning of anti-viral drugs as therapy for cervical cancer, Pharmacol. Rep. 68 (5) (2016) 983–989.
[55] A.L. Oei, et al., Hyperthermia selectively targets human papillomavirus in cervical tumors via p53-dependent apoptosis, Cancer Res. 75 (23) (2015) 5120–5129.
[56] B.M. Hougardy, et al., Proteasome inhibitor MG132 sensitizes HPV-positive human cervical cancer cells to rhTRAIL-induced apoptosis, Int. J. Cancer 118 (8) (2006) 1892–1900.
[57] Y. Yang, et al., Targeting the ubiquitin-proteasome system for cancer therapy, Cancer Sci. 100 (1) (2009) 24–28.

[58] D.C. Birle, D.W. Hedley, Suppression of the hypoxia-inducible factor-1 response in cervical carcinoma xenografts by proteasome inhibitors, Cancer Res. 67 (4) (2007) 1735–1743.
[59] S. Kamer, Q. Ren, A.P. Dicker, Differential radiation sensitization of human cervical cancer cell lines by the proteasome inhibitor velcade (bortezomib, PS-341), Arch. Gynecol. Obstet. 279 (1) (2009) 41–46.
[60] T.J. Pugh, et al., Phase I trial of bortezomib and concurrent external beam radiation in patients with advanced solid malignancies, Int. J. Radiat. Oncol. Biol. Phys. 78 (2) (2010) 521–526.
[61] B.R. Cullen, RNA interference: antiviral defense and genetic tool, Nat. Immunol. 3 (7) (2002) 597–599.
[62] P.A. Sharp, RNA interference—2001, Genes Dev. 15 (5) (2001) 485–490.
[63] A.H.S. Hall, K.A. Alexander, RNA interference of human papillomavirus type 18 E6 and E7 induces senescence in HeLa cells, J.Virol. 77 (10) (2003) 6066–6069.
[64] M. Aarthy, U. Panwar, S.K. Singh, Structural dynamic studies on identification of EGCG analogues for the inhibition of human papillomavirus E7, Sci. Rep. 10 (1) (2020) 8661.
[65] N.S.L. Yeo-Teh, Y. Ito, S. Jha, High-risk human Papillomaviral oncogenes E6 and E7 target key cellular pathways to achieve oncogenesis, Int. J. Mol. Sci. 19 (6) (2018) 1706.
[66] E.-K. Yim, J.-S. Park, The role of HPV E6 and E7 oncoproteins in HPV-associated cervical carcinogenesis, Cancer Res. Treat. 37 (6) (2005) 319–324.
[67] A. Goel, A.B. Kunnumakkara, B.B. Aggarwal, Curcumin as "Curecumin": from kitchen to clinic, Biochem. Pharmacol. 75 (4) (2008) 787–809.
[68] P. Anand, et al., Curcumin and cancer: an "old-age" disease with an "age-old" solution, Cancer Lett. 267 (1) (2008) 133–164.
[69] S. Aggarwal, et al., Curcumin (diferuloylmethane) down-regulates expression of cell proliferation and antiapoptotic and metastatic gene products through suppression of IkappaBalpha kinase and Akt activation, Mol. Pharmacol. 69 (1) (2006) 195–206.
[70] R.L. Hong, W.H. Spohn, M.C. Hung, Curcumin inhibits tyrosine kinase activity of p185neu and also depletes p185neu, Clin. Cancer Res. 5 (7) (1999) 1884–1891.
[71] A.L. Cheng, et al., Phase I clinical trial of curcumin, a chemopreventive agent, in patients with high-risk or pre-malignant lesions, Anticancer Res 21 (4b) (2001) 2895–2900.
[72] C.S. Divya, M.R. Pillai, Antitumor action of curcumin in human papillomavirus associated cells involves downregulation of viral oncogenes, prevention of NFkB and AP-1 translocation, and modulation of apoptosis, Mol. Carcinog. 45 (5) (2006) 320–332.
[73] C.L. Kuo, C.W. Chi, T.Y. Liu, The anti-inflammatory potential of berberine in vitro and in vivo, Cancer Lett. 203 (2) (2004) 127–137.
[74] S.K. Mantena, S.D. Sharma, S.K. Katiyar, Berberine, a natural product, induces G1-phase cell cycle arrest and caspase-3-dependent apoptosis in human prostate carcinoma cells, Mol. Cancer Ther. 5 (2) (2006) 296–308.
[75] P.L. Peng, et al., Inhibitory effect of berberine on the invasion of human lung cancer cells via decreased productions of urokinase-plasminogen activator and matrix metalloproteinase-2, Toxicol. Appl. Pharmacol. 214 (1) (2006) 8–15.
[76] S. Mahata, et al., Berberine modulates AP-1 activity to suppress HPV transcription and downstream signaling to induce growth arrest and apoptosis in cervical cancer cells, Mol. Cancer 10 (2011) 39.
[77] P. Paul, et al., Novel action modality of the diterpenoid anisomelic acid causes depletion of E6 and E7 viral oncoproteins in HPV-transformed cervical carcinoma cells, Biochem. Pharmacol. 89 (2) (2014) 171–184.
[78] W. Zhang, et al., Marine Streptomyces sp. derived antimycin analogues suppress HeLa cells via depletion HPV E6/E7 mediated by ROS-dependent ubiquitin–proteasome system, Sci. Rep. 7 (1) (2017) 42180.

[79] A. Ghanbari, et al., Biological activities of Ficus carica latex for potential therapeutics in human papillomavirus (HPV) related cervical cancers, Sci. Rep. 9 (1) (2019) 1013.

[80] J.F. Sah, et al., Epigallocatechin-3-gallate inhibits epidermal growth factor receptor signaling pathway. Evidence for direct inhibition of ERK1/2 and AKT kinases, J. Biol. Chem. 279 (13) (2004) 12755–12762.

[81] M.S. Longworth, L.A. Laimins, The binding of histone deacetylases and the integrity of zinc finger-like motifs of the E7 protein are essential for the life cycle of human papillomavirus type 31, J. Virol. 78 (7) (2004) 3533–3541.

[82] P. Marks, et al., Histone deacetylases and cancer: causes and therapies, Nat. Rev. Cancer 1 (3) (2001) 194–202.

[83] P. Finzer, et al., Growth arrest of HPV-positive cells after histone deacetylase inhibition is independent of E6/E7 oncogene expression, Virology 304 (2) (2002) 265–273.

[84] P. Finzer, et al., Inhibitors of histone deacetylase arrest cell cycle and induce apoptosis in cervical carcinoma cells circumventing human papillomavirus oncogene expression, Oncogene 20 (35) (2001) 4768–4776.

[85] P. Finzer, et al., HDAC inhibitors trigger apoptosis in HPV-positive cells by inducing the E2F-p73 pathway, Oncogene 23 (28) (2004) 4807–4817.

[86] N. Sima, et al., RNA interference against HPV16 E7 oncogene leads to viral E6 and E7 suppression in cervical cancer cells and apoptosis via upregulation of Rb and p53, Apoptosis 13 (2) (2008) 273–281.

[87] D.-Y. Fu, et al., Peptidic inhibitors for in vitro pentamer formation of human papillomavirus capsid protein l1, ACS Med. Chem. Lett. 6 (4) (2015) 381–385.

[88] D.D. Zheng, et al., Efficient inhibition of human papillomavirus 16 L1 pentamer formation by a carboxylatopillarene and a p-sulfonatocalixarene, Chem. Commun. (Camb.) 50 (24) (2014) 3201–3203.

[89] Y. Zhao, et al., Doxycycline inhibits proliferation and induces apoptosis of both human papillomavirus positive and negative cervical cancer cell lines, Can. J. Physiol. Pharmacol. 94 (5) (2016) 526–533.

[90] K.L. Borden, B. Culjkovic-Kraljacic, Ribavirin as an anti-cancer therapy: acute myeloid leukemia and beyond? Leuk. Lymphoma 51 (10) (2010) 1805–1815.

[91] G. Andrei, E. De Clercq, R. Snoeck, Drug targets in cytomegalovirus infection, Infect. Disord. Drug Targets 9 (2) (2009) 201–222.

[92] R. Snoeck, et al., Cidofovir, a new approach for the treatment of cervix intraepithelial neoplasia grade III (CIN III), J. Med. Virol. 60 (2) (2000) 205–209.

[93] S. Park, et al., HIV-1 protease inhibitors slow HPV16-driven cell proliferation through targeted depletion of viral E6 and E7 Oncoproteins, Cancer 13 (5) (2021) 949.

[94] L. Hampson, H.C. Kitchener, I.N. Hampson, Specific HIV protease inhibitors inhibit the ability of HPV16 E6 to degrade p53 and selectively kill E6-dependent cervical carcinoma cells in vitro, Antivir. Ther. 11 (6) (2006) 813–825.

[95] A.D. Soriano-Hernandez, et al., Anti-inflammatory drugs and uterine cervical cancer cells: antineoplastic effect of meclofenamic acid, Oncol. Lett. 10 (4) (2015) 2574–2578.

[96] S.H. Kim, et al., GADD153 mediates celecoxib-induced apoptosis in cervical cancer cells, Carcinogenesis 28 (1) (2007) 223–231.

[97] G. Friel, et al., Aspirin and acetaminophen use and the risk of cervical cancer, J. Low. Genit. Tract Dis. 19 (3) (2015) 189–193.

[98] A. Mondal, U. Chatterji, Artemisinin represses telomerase subunits and induces apoptosis in HPV-39 infected human cervical Cancer cells, J. Cell. Biochem. 116 (9) (2015) 1968–1981.

[99] G.L. Disbrow, et al., Dihydroartemisinin is cytotoxic to papillomavirus-expressing epithelial cells in vitro and in vivo, Cancer Res. 65 (23) (2005) 10854–10861.

[100] S. D'Alessandro, et al., The use of antimalarial drugs against viral infection, Microorganisms 8 (1) (2020) 85.

[101] A. Kumar, E. Rathi, S. Kini, Drug repurposing approach for the identification and designing of potential E6 inhibitors against cervical cancer: an in silico investigation, Struct. Chem. 31 (2020).
[102] S. Wang, et al., HPV E6 induces eIF4E transcription to promote the proliferation and migration of cervical cancer, FEBS Lett. 587 (6) (2013) 690–697.
[103] I. Chopra, P.M. Hawkey, M. Hinton, Tetracyclines, molecular and clinical aspects, J. Antimicrob. Chemother. 29 (3) (1992) 245–277.
[104] L.M. Coussens, Z. Werb, Inflammation and cancer, Nature 420 (6917) (2002) 860–867.
[105] X. Xin, et al., Targeting COX-2 and EP4 to control tumor growth, angiogenesis, lymphangiogenesis and metastasis to the lungs and lymph nodes in a breast cancer model, Lab. Invest. 92 (8) (2012) 1115–1128.
[106] D. Davila-Gonzalez, J.C. Chang, T.R. Billiar, NO and COX2: dual targeting for aggressive cancers, Proc. Natl. Acad. Sci. 114 (52) (2017) 13591–13593.
[107] K. Münger, et al., Interactions of HPV E6 and E7 oncoproteins with tumour suppressor gene products, Cancer Surv. 12 (1992) 197–217.
[108] S. Sanami, et al., Design of a multi-epitope vaccine against cervical cancer using immunoinformatics approaches, Sci. Rep. 11 (1) (2021) 12397.
[109] J.J.P. Alves, et al., Th17 response in patients with cervical cancer (review), Oncol. Lett. 16 (5) (2018) 6215–6227.
[110] C.B. Buck, P.M. Day, B.L. Trus, The papillomavirus major capsid protein L1, Virology 445 (1–2) (2013) 169–174.
[111] S. DiGiuseppe, et al., The nuclear retention signal of HPV16 L2 protein is essential for incoming viral genome to transverse the trans-Golgi network, Virology 458-459 (2014) 93–105.
[112] A. Namvar, et al., In silico/in vivo analysis of high-risk papillomavirus L1 and L2 conserved sequences for development of cross-subtype prophylactic vaccine, Sci. Rep. 9 (1) (2019) 15225.
[113] N. Ghorban Hosseini, et al., In silico analysis of L1/L2 sequences of human papillomaviruses: implication for universal vaccine design, Viral Immunol. 30 (3) (2017) 210–223.
[114] T.T. Ashburn, K.B. Thor, Drug repositioning: identifying and developing new uses for existing drugs, Nat. Rev. Drug Discov. 3 (8) (2004) 673–683.
[115] S. Pushpakom, et al., Drug repurposing: progress, challenges and recommendations, Nat. Rev. Drug Discov. 18 (1) (2019) 41–58.
[116] U. Sahu, et al., Interleukin-17—a multifaceted cytokine in viral infections, J. Cell. Physiol. (2021) 1–2.
[117] U. Sahu, P. Khare, Role of interleukin-17 in human papillomavirus infection and associated malignancies, Microb. Pathog. (2021) 105294.

CHAPTER 9

Recent advances and strategies in vaccine development against HPV

Anoop Kumar[a] and Praveen Rai[b]
[a]National Institute of Biologicals (NIB), Ministry of Health & Family Welfare, Government of India, Noida, Uttar Pradesh, India
[b]Division of Infectious Diseases, Nitte (Deemed to be University), Nitte University Centre for Science Education & Research (NUCSER), Deralakatte, Karnataka, India

1. Introduction

Human papillomaviruses (HPV) are ubiquitous DNA viruses belonging to family *papillomaviridae* and responsible for causing anogenital cancer and warts [1,2]. More than 118 HPV types have been defined to date. Some of which are high-risk HPV (HR-HPVs) types (HPV16, 18, 31, 33, 35, 39, 45, 51, 52, 56, 58, 59, 68, 73, and 82) that are linked with genital and other epithelial cancers, and others are low-risk HPV (LR-HPVs) types (HPV6, 11, 40, 42, 43, 44, 54, 61, 70, 72, 81, and CP6108a) which are responsible for benign tumors and genital warts [1–3]. It is a well-known fact that sexually transmitted HPV infection is the etiological agent in most cases of cervical cancer (CC) [2,4]. Identification of an etiological infectious agent allows preventing hyperplastic, papillomatous, and verrucous squamous cell lesions in the skin and at various mucosal sites in various hosts, including humans, and exploit the cellular machinery for their purposes [5].

HPV infects various body sites such as anogenital tract, nasal cavity, urethra, skin, tracheobronchial mucosa, larynx, conjunctiva, and esophagus [6,7]. Worldwide, it is the most common sexually transmitted infection (STI), causing approximately 4.5% of all cancer cases [8] and genital warts. Infection with HR-HPV is not only allied with CC but also with a generous number of oral, esophageal, and other anogenital cancers, namely cancers of vulva, vagina, penis, and anus [9–12]. Infection with LR-HP (mainly HPV-6 and HPV-11) causes genital warts or benign tumors such as common warts, condylomas, intraepithelial neoplasias, anogenital, oropharyngeal papillomas, keratoacanthomas, and other types of hyperkeratoses. These LR-HPV types can also cause recurrent respiratory papillomatosis

Immunopathology, Diagnosis, and Treatment of HPV-induced Malignancies
https://doi.org/10.1016/B978-0-323-91797-1.00011-9

(RRP), a rare but life-threatening disease that can require patients to endure multiple surgeries each year to remove warts that block their upper airways [13].

A very high frequency (98%) of HR-HPV infection is observed in women with CC that strongly supports their role in cervical carcinogenesis, with utmost dominant HR-HPV, i.e., HPV-16, followed by other HR-HPVs types 18, 31, 33, and 45 [14–16]. HR-HPV type, i.e., HPV-16/18 is prospective to be accountable for about 80% of invasive CC comprising cervical intraepithelial neoplasia (CIN 3), vulvar intraepithelial neoplasia (VIN) 2/3, and for 50% of CIN 2 lesions. In addition to HPV infection, there are several other cofactors, including host genetic and cellular factors, HPV variants, immunodeficiency, viral integration, and viral load, that may be necessary for the progression of cervical lesions to cancer [17,18] (Fig. 1).

The host immune system identifies the foreign agents (antigens) that conquer the body and fight with them by using innate and adaptive, but these approaches are occasionally ineffective against HPV. HPV uses many strategies to escape from the host immune system [19]. The HPV mainly

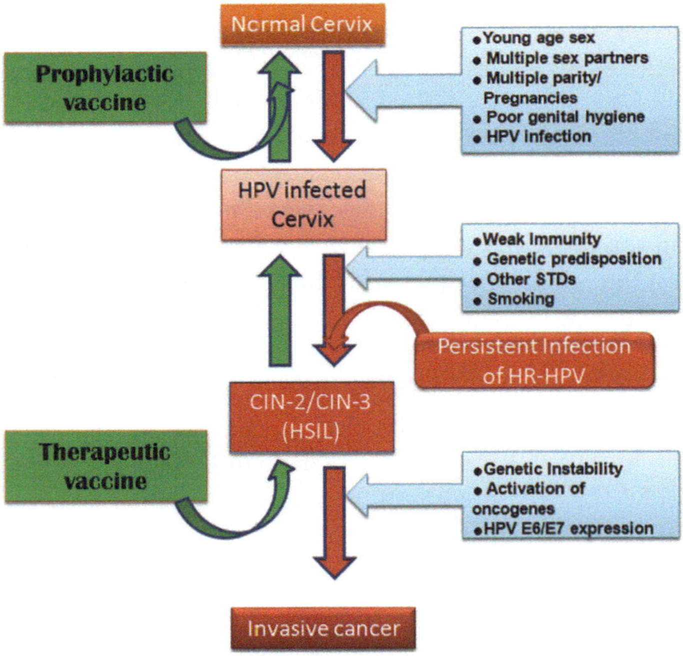

Fig. 1 The natural history and biological behavior of HPV infection and cervical lesions along with the possible role of other factors.

infects and divides in keratinocytes; these keratinocytes have a short life span due to which HPV need not destroy the cell due to which the inflammation/immune response is not triggered [19]. Additionally, these keratinocytes are distant from immune cells. The expression of various immune genes like interferon are downregulated by HPV [20]. Therefore, an effective vaccine based on chimeric VLPs, nucleic acid, or live vector vaccine can provide an effective immune response.

2. Requirement and type of HPV vaccine

Very robust biological, clinical, and epidemiologic data suggested that HPVs are the etiological agent for the development of the cancer of the cervix [21]. This contagious etiology offers a prospect to prevent a key root of HPV-mediated CC deaths in women through vaccination. The need to prevent or cure genital HPV infection by using vaccination has directed investigators to work on various tactics for the development of candidates for the HPV vaccines [22]. The vaccination prevents and/or treats the benign and malignant states caused by infection of HPV through the immunogenic effect in a cost-effective manner [22–24]. The candidate vaccine of HPV should reduce the incidence of CC as well as its precursor lesions and other cancers, like oropharyngeal cancer, and head and neck cancers, caused by the infection of HPV [25,26]. Different HPV genotypes have been witnessed in diverse geographic areas [27]. To project and assess the health impact of HPV infections and the prospective benefits of the execution of future vaccines, it is essential to know the distribution of HPV oncogenic types in diverse geographic regions and different populations.

2.1 Prophylactic HPV vaccines

Prophylactic vaccines prevent infection by stimulating antibody-mediated immunity which is proficient in identifying and inactivating the virus before it can infect the host cells [28]. The properties of an ideal prophylactic HPV vaccine are that, first, it should be safe for all individuals to whom it is administered as they would be mainly normal individuals having no HPV infection and would not be expected to develop HPV infection leading to cancer. Second, the immunogenic effect of the vaccine should be long-lasting. The major goal of the HPV vaccine should be its effectiveness against the most frequent HPV type identified in CC [29]. Worldwide, most CC cases (more than 50%) are infected with HPV 16 compared to all types of HPV [30].

2.2 Therapeutic vaccines

Therapeutic vaccines help in the treatment of women who are already infected with HPV. Therapeutic vaccines could help in the prevention of the progression of the low-grade disease and cause the regression of the existing lesions. The therapeutic vaccines have the potential to control the spread and recurrence of metastatic cancer after treatment [31,32]. An effective therapeutic HPV vaccine must have cell-mediated immunity because once the virus has been incorporated into the host cells, it is difficult for antibodies to reach and eliminate the virus from the host cell [31,33,34].

3. Candidate for the HPV vaccine development

HPVs cannot be grown in the laboratory as a source of antigen for serological tests and conventional killed vaccine development, and they do not cause diseases in animals, so vaccine development is difficult. Therefore, HPV vaccines are currently under development, and the developed vaccines employ genetic engineering technology. The desire to prevent or treat genital HPV infection through immunization has led investigators to employ several strategies to develop candidate HPV vaccines [22].

A key innovation in the progress of HPV vaccine research came with the discovery that the capsid proteins L1 and L2 (or L1 alone) of HPV can self-assemble into VLPs when expressed in appropriate host cells. VLPs closely resemble native HPV particles. VLPs are not only immunogenic and safe but also able to induce strong cell-mediated and humoral immune responses [35] in a controlled trial of the HPV16 vaccine, and nearly 100% effectiveness was achieved.

3.1 VLPs prophylactic candidates

For the development of HPV vaccine, VLPs is an attractive candidate for fighting viral infections as it has highly immunogenic repetitive epitopes found on virions surface, yet VLPs are devoid of the potentially harmful viral genomes. Various in vitro and in vivo studies of papillomavirus VLPs, which consist of L1 major virion protein, moved this as a lead candidate for HPV vaccines [36]. The main requirement for prophylactic HPV vaccines is to induce neutralizing antibodies against natural structural viral capsid proteins to prevent virus entry into the host cell. The prerequisite to obtaining this effect is that the immunogen should possess the natural structure to induce the conformation-dependent antibodies, and the route of immunization should favor the activation of mucosal immunity.

3.2 E6/E7 oncoproteins as therapeutic candidates

The prophylactic HPV vaccine can prevent only when the girls have not been infected with HPV; however, therapeutic vaccines target cell-mediated immune response to clear existing infection [37,38]. Therefore, a therapeutic vaccine expressing E6 and E7 oncoproteins of HPV may be an ideal target [39]. These early proteins help in maintaining the malignant phenotype of the HPV-infected cells [39].

4. Licensed vaccines against HPV

Currently, the following three licensed prophylactic HPV vaccines (Table 1) are available in various countries throughout the world:

1. Gardasil, a quadrivalent HPV vaccine that targets HPV6, HPV11, HPV16, and HPV18.
2. Cervarix, a bivalent HPV vaccine that targets HPV16, and HPV18.
3. Gardasil 9, a nonavalent HPV vaccine that targets HPV6, 11, 16, 18, 31, 33, 45, 52, and 58.

The three HPV vaccines are greatly efficacious in preventing HPV infection against major HR-HPV types, i.e., 16 and 18, which are accountable for almost 80% of CC cases and preventing precancerous cervical lesions worldwide. Gardasil 9, a nonavalent HPV vaccine, delivers additional safeguard against other HR-HPV types such as HPV31, 33, 45, 52, and 58.

These three HPV vaccines showed a lot of promise, but they may not be appropriate for the low-resource developing countries. Vaccines for the resource-limited countries should have a low cost of production, have a longer shelf life, single-dose delivery, long-lasting immunity, have both prophylactic and therapeutic value, should be stable at room temperature, and could incorporate other oncogenic HPVs [40–43]. Administration of vaccines orally or via nasal spray rather than via injection may eliminate the need for sterile needles and highly trained providers [26,42].

5. Development of the future HPV vaccine

5.1 L1 capsomere-based HPV vaccine

Instead of making complex VLPs, a cheaper, equally effective, and thermostable alternative would be the capsomere-based vaccine, where only five instead of 360 L1 monomers are required to assemble a thermostable pentavalent capsomere of HPV. It could be a potent alternative cost-effective vaccine candidate. This can be manufactured in bacteria

Table 1 Details of the three licensed HPV vaccines.

Category	Gardasil	Cervarix	Gardasil 9
Manufacturer	Merck & Co., Inc.	GlaxoSmithKline Biologicals	Merck & Co., Inc.
Vaccines types	L1 VLP (quadrivalent)	L1 VLP (bivalent)	L1 VLP (nonavalent)
Produced In	*Saccharomyces cerevisiae*	Baculovirus	*Saccharomyces cerevisiae*
Types of HPVs	6, 11, 16, and 18	16 and 18	6, 11, 16, 18, 31, 33, 45, 52, and 58
Injection schedule	Three injections at months 0, 2, and 6	Three injections at months 0, 1, and 6	Two or three injections at months 0, 2, and 6
Adjuvant AS04	Aluminum hydroxyphosphate sulphate	Alum + adjuvant MPL (Monophosphoryl lipid A)	Amorphous aluminum hydroxyphosphate sulphate (AAH500)
Target Population	9–25 years (male and female)	9–26 years (female)	9–45 years (male and female)
FDA approval	2006	2007	2014
Immunogenicity	100% sero-conversion	100% sero-conversion	98% sero-conversion
Efficacy in preventing 100% persistent infection	89%	100%	>96%
Efficacy in preventing 100% CIN lesion	100%	100%	98.2%
Cross protection	Not known	HPV 45 and 31	Not known

like *Salmonella typhimurium* and *Escherichia coli*, further reducing the cost. Dell et al. [41] produced pentameric capsomeres of HPV 16 L1 in *E. coli* and showed the induction to produce neutralizing antibodies in animal models [41]. The pentameric capsomeres represent the greatly stable substitute to VLPs, and they can be cheaper to produce and easily expressed in *E. coli*. To produce cheaper vaccines, researchers trying to express the VLP L1 and other viral proteins in bacteria or plants to reduce the cost of production of the vaccine. If HPV16 L1 VLP could be produced in transgenic plants or transiently expressed edible plants, their fruits and vegetables could theoretically serve as an edible vaccine. They can be delivered orally as it is easily digestible and immunogenic. Some studies have shown that HPV VLPs administered orally to mice produce type-specific antibody responses in serum and genital mucosal secretions [44]. But for the induction of high antibody response by injecting orally as compared to other means of injection, e.g., parenteral injection, more VLPs are required [42]. Several additional attempts are being made to decrease the cost ofVLP vaccine delivery by making needle-free mucosal spray or oral delivery [40]. A phase-II clinical trial of HPV 16/18 L1 capsomere-based vaccine-derived in an *E. coli* was conducted (NCT 01355823) [45].

5.2 L2-based prophylactic HPV vaccines

Inherent limitations of L1 VLPs-based vaccines of being expensive and having a narrow spectrum range can be addressed by designing an alternative form of the vaccine. Bacteriophage PP7- and *Lactobacillus caseii*-based L2 VLPs are two L2 prophylactic vaccines against HPV. These vaccines are not only cheaper but have a broader spectrum range due to conserved amino acid sequences in most of HR-HPVs. Antigens against these conserved regions can induce neutralizing antibodies against a wide spectrum of HPV types including most oncogenic types. Using *E. coli* as a host cell for mass production of the vaccine may further reduce the cost [46,47].

5.3 Chimeric VLP vaccines

Chimeric VLP vaccine, which constitutes both the L1 and L2 protein components, has good immunogenicity conferred by L1 component and broad-spectrum conferred by L2 proteins. It would combine the advantages of both, i.e., resulting in a highly immunogenic vaccine with wider coverage. Boxus et al. [48] used AS04 adjuvanted based on VLP chimeras of L1 along with one or two L2 epitopes, which were inserted within the DE loop and/or C terminus of L1. This chimeric vaccine construct of L1/L2 VLP induced good immune responses and protection in animal models against different HPVs (HPV6, 11, 16, 31, 35, 39, 45, 58, and 59 as quasivirions or pseudovirions) [48].

5.4 Vector-based vaccines

The major advantage of using live vector-based vaccines for introducing antigen is their property to replicate within the host cell, thereby spreading the antigen efficiently within the host. Despite strong immunogenicity, the efficiency of vaccines is hampered by the development of neutralizing antibodies which will reduce the potency of repeated immunization. Other side -effects could be immunosuppressive factors and potential toxicity elicited by the vector and its components. Attenuated bacterial vectors such as *Listeria monocytogenes*, *Salmonella*, *Lactococcus lactis* [49,50], *Lactobacillus plantarum* [51], and *Bacillus Calmette Guérin* are currently being used to deliver antigen to dendritic cells (DCs), which further activate T cells inducing a strong cell-mediated immune response. *Listeria monocytogenes* derived ADXS11-001, an HPV vaccine, successfully elicits immune responses against oncoprotein E7 in HPV-related cancer like cervical, oropharyngeal, and anal cancers. Recently, a phase I/II clinical trial was conducted using the listeria-based HPV therapeutic vaccine targeting the HPV-E7 antigen in patients with cancer of cervical [50].

The vaccinia virus, a viral vector, has been successful in augmenting host immune response due to its incredible efficiency of infection. HPV vaccine based on the vaccinia virus vector has shown good potential for regression of diseases in clinical trials. A modified vaccinia Ankara (MVA) vector-based vaccine having E6/E7 of HPV16 and IL-2 showed a clinical response in 48% of the with CIN2/3 [52].

5.5 Protein/peptide-based vaccine

Direct administration of HPV antigens is presented by DCs to activate antigen-specific T cells. They are safe, stable, and easy to produce. They are major histocompatibility complex (MHC) making it critical to recognize precise immunogenic epitopes of HPV antigens for a vaccine owing to the polymorphic nature of HLA molecules, making it ineffective in a population with huge variation in the haplotype. These protein/peptide-based vaccines are safe, stable, and easy to produce but require immunostimulating molecules like adjuvant to increase the immunogenicity [53].

The other essential concern for the development of the HPV vaccine based on protein/peptide is the heterogeneity of human leukocyte antigen

(HLA), which causes altered immune responses to protein/peptide-based vaccine [54]. So, the generation of the good cancer cell-specific CTLs response requires long, overlapping peptides require to escape HLA [55]. The tumor regression in the animal model showed a promising response by E6 and E7 peptide-based HPV vaccines [56].

5.6 Nucleic acid-based vaccines

5.6.1 DNA-based HPV vaccines

DNA vaccines are plasmid-based vaccines having the specific antigen gene (small HPV gene) inserted in the plasmid. After the plasmid construction, they are produced at a large scale and then purified to make them ready to inject into the vaccine recipients by appropriate methods like intramuscularly in saline solutions, encapsulated in microparticles, or by "gene gun" methods [34,57,58]. The cells take up the injected plasmid DNA and then produce the selective cell-mediated immune response and antibody-mediated immunity for a longer period [33].

The development of DNA-based vaccines has occurred as an imperative strategy for the development of the HPV vaccine. The DNA vaccine is a plasmid-based vaccine having antigenic genes of interest, like major (L1) and/or minor capsid (L2) (as a prophylactic vaccine) and E6/E7 oncogenic proteins (as a therapeutic vaccine). Plasmid DNA having the gene of the interest is injected into the host and stimulates a specific cell-mediated immune response against the injected antigen. Several reports are showing that DNA vaccine encoding HPV16, E6, and E7 genes can induce E6- and E7-specific cytotoxic T-lymphocyte response and can efficiently control HPV-associated lesions in animal models. Several other strategies are under process to develop an effective immune response of DNA-based vaccines against HPV.

Matijevic et al. [59] show the vaccinated with a poly(lactide-*co*-glycolide) captured DNA plasmid stating antigenic regions of HPV16 and HPV18 result in a rise in the precursor frequency of T-cells that respond to epitopes from HPV16, 18, 6, and 11 [59]. Haigh et al. [60] used an HBsAg-based DNA vaccine as a carrier to induce a specific response for the co-encoded tumor antigens [59]. They used the fusion of Hepatitis B surface antigen and delivered by DNA vaccination to elicit CTL responses to HPV oncoproteins allied with tumor protection and the advancement for the development of a therapeutic vaccine for HPV-associated squamous carcinomas. Zhou et al. [61] showed the immunogenic response of recombinant vaccinia virus expressing bivalent E7/E6 fusion proteins from HPV16 and 18 vaccine construct in vaccinated mice and rhesus monkeys [61].

5.6.2 RNA replicon vaccines

RNA replicons comprise of naked RNA molecules that reproduce within the transfected cells. Alphaviruses like Semliki Forest virus [62], Sindbis virus [63,64], and Venezuelan equine encephalitis virus (VEE) [65] are good sources of RNA replicon. They have an outstanding feature of replicating in widespread cell types leading to a sustained level of antigen expression in cells, which establishes them as good candidate development of cancer vaccine. However, there is a potential risk of their integration into the host genome. A new version of DNA launched RNA replicon, termed "suicidal" DNA, has been used for the development of the HPV vaccine, which is being studied in preclinical models [66,67]. "Suicidal DNA" is copied into RNA in the transfected cell-like DCs and offers an efficient and stable way of tumor antigen expression. The apoptosis of transfected cells is averted by linking the suicidal DNA encoding the antigen like E7 to antiapoptotic proteins, for instance BCL-xL, and causes expressively higher antigen-specific $CD8^{+}$ T-cell immune responses and enhanced antitumor effects than DNA vaccine alone [67]. Multiple variations have been tried out to improve the efficacy of suicidal DNA, like employing replication-defective, self-limiting replicon particles using a safe packaging cell line which has the advantage of efficient gene delivery and large-scale production. The vigor of HPV RNA replicon vaccines can be boosted by using intracellular targeting and spreading strategies [68].

5.7 Whole cell-based HPV vaccines

5.7.1 DC-based vaccines

DCs loaded with viral antigens represent an acceptable approach for HPV therapeutic vaccines. DCs prepared ex vivo by the physical loading of MHC-I and MHC-II molecules are pulsed with antigen in the form of peptides or proteins or transfected with HPV antigen encoded in DNA or RNA are delivered via recombinant adenoviral vectors carrying HPV antigens like E7 [69] or DCs are infected with tumor antigen-expressing vaccinia virus [70]. They can be injected via the intramuscular, subcutaneous, or intravenous route. However, multiple limitations are associated with DC-based vaccines, like designing individualized vaccines as a cumbersome task, culturing method used to formulate DCs to vary the vaccine quality, and a lack of standard criteria for evaluating the vaccines. Several approaches have been deployed to home pulsed DCs to the lymphoid organs effectively to stimulate a strong immune response [69]. The efficacy of this vaccine can be further augmented by prolonging DC survival and improving T-cell priming by DCs.

It can be achieved by either transfecting antigen-loaded DCs with small interfering RNA (siRNA) aiming the proapoptotic proteins, Bak and Bax, leads to downregulation of Bak and Bax protein expression in the DCs, which consequence in amplified resistance of the DCs to T-cell-mediated apoptosis, thus enhancing antigen like E7-specific immune responses and antitumor effects [71]. Santin et al. [72] showed that injecting a DCs pulsed with recombinant E7 of HPV-16 or -18 was shown to persuade E7-specific $CD4^+$ T cell immune and E7-specific $CD8^+$ T-cell responses in patients with cancer of cervix refractory to standard cures [72].

5.7.2 Tumor cell-based vaccines

The second approach for the whole cell-based immunization used tumor cells in HPV vaccines. Vaccines based on whole tumor cells result from the tumor cells patient which is cumbersome and a laborious method that needs a substantial amount of tumor tissue. Discrepancies in the purity and potency of these vaccines constrain their efficacy and limit the scope for the development of HPV vaccine. The advantage of using this vaccine is that it precludes the need for antigen preselection.

The use of autologous tumor cells may not do miracles since the host immune response has already failed against antigens expressed by the tumor. Anyhow, they can be influenced ex vivo to express immunomodulatory proteins, for example, cytokines, to increase their immunogenicity. Cytokine genes used in HPV transformed tumor vaccines include IL-2 [73], IL-12 [70,74], and granulocyte macrophage colony-stimulating factor (GMCSF) [75]. Recently, vaccination of mice with GMCSF expressing E7-positive tumor cells led to an E7-specific CTL response and potent antitumor effects against E7-expressing tumors in vivo [75].

5.8 Epitope-based vaccine

For the development of the cost-effective and immunogenic vaccine, it is vital to identify the potential epitopic regions which can elicit a potent immune response against the pathogen. Identification of the potential T- and B-cells can be identified using immunoinformatics, a branch of bioinformatics [76]. The prediction of the potential epitopes helps in shorting the peptide used for the development of the vaccine cost-effectively as it can reduce the screening of nonimmunogenic epitopes and reduce the cost of the vaccine development.

6. The future of HPV vaccination

Chimeric vaccine having both prophylactic and therapeutic efficacies with strategies that boost long-lasting host immune response so as to lower the CC burden worldwide is an inescapable necessity. In silico tools, harmonizing with enhancements in recombinant/chimeric DNA (rDNA) technology and information on the immune response of the host and the genetic background of the HPV will contribute to the development of new vaccines in the future.

7. Conclusion

Knowing the target that allows preventing HPV-mediated cancer and other diseases using the existing HPV prophylactic vaccine; however, they lack therapeutic potential. The available recent advancement in immunobiology gives high optimism for the development of the highly effective therapeutic vaccine for HPV in a cost-effective manner. The development of the cost-effective HPV vaccine having both prophylactic and therapeutic potential will benefit the reduction of HPV-mediated disease worldwide.

References

[1] H. Kelly, et al., A systematic review and meta-analysis of studies evaluating the performance of point-of-care tests for human papillomavirus screening, Sex. Transm. Infect. 93 (S4) (2017) S36–S45.

[2] H. zur Hausen, Human genital cancer: synergism between two virus infections or synergism between a virus infection and initiating events? Lancet 2 (8312) (1982) 1370–1372.

[3] N. Munoz, et al., Epidemiologic classification of human papillomavirus types associated with cervical cancer, N. Engl. J. Med. 348 (6) (2003) 518–527.

[4] H. Zur Hausen, Papillomavirus infections—a major cause of human cancers, Biochim. Biophys. Acta 1288 (2) (1996) F55–F78.

[5] D.R. Lowy, et al., In vitro tumorigenic transformation by a defined sub-genomic fragment of bovine papilloma virus DNA, Nature 287 (5777) (1980) 72–74.

[6] B.C. Das, et al., Prospects and prejudices of human papillomavirus vaccines in India, Vaccine 26 (22) (2008) 2669–2679.

[7] K.J. Syrjanen, HPV infections and oesophageal cancer, J. Clin. Pathol. 55 (10) (2002) 721–728.

[8] C. de Martel, et al., Worldwide burden of cancer attributable to HPV by site, country and HPV type, Int. J. Cancer 141 (4) (2017) 664–670.

[9] S. de Sanjose, et al., Population-based e-records to evaluate HPV triage of screen-detected atypical squamous cervical lesions in Catalonia, Spain, 2010-15, PLoS One 13 (11) (2018), e0207812.

[10] T. Hoory, et al., Molecular epidemiology of human papillomavirus, J. Formos. Med. Assoc. 107 (3) (2008) 198–217.

[11] H. zur Hausen, Papillomavirus in anogenital cancer: the dilemma of epidemiologic approaches, J. Natl. Cancer Inst. 81 (22) (1989) 1680–1682.
[12] H. zur Hausen, Papillomaviruses in anogenital cancer as a model to understand the role of viruses in human cancers, Cancer Res. 49 (17) (1989) 4677–4681.
[13] M. Carifi, et al., Recurrent respiratory papillomatosis: current and future perspectives, Ther. Clin. Risk Manag. 11 (2015) 731–738.
[14] E.M. de Villiers, Human pathogenic papillomavirus types: an update, Curr. Top. Microbiol. Immunol. 186 (1994) 1–12.
[15] L.A. Laimins, The biology of human papillomaviruses: from warts to cancer, Infect. Agents Dis. 2 (2) (1993) 74–86.
[16] H. zur Hausen, E.M. de Villiers, Human papillomaviruses, Annu. Rev. Microbiol. 48 (1994) 427–447.
[17] U. Sahu, et al., Interleukin-17—a multifaceted cytokine in viral infections, J. Cell. Physiol. (2021) 1–2.
[18] U. Sahu, P. Khare, Role of interleukin-17 in human papillomavirus infection and associated malignancies, Microb. Pathog. (2021), 105294.
[19] C. Zhou, Z.K. Tuong, I.H. Frazer, Papillomavirus immune evasion strategies target the infected cell and the local immune system, Front. Oncol. 9 (2019) 682.
[20] M. Stanley, Chapter 17: genital human papillomavirus infections—current and prospective therapies, J. Natl. Cancer Inst. Monogr. 31 (2003) 117–124.
[21] IARC Working Group on the Evaluation of Carcinogenic Risks to Humans, Human papillomaviruses, IARC Monogr. Eval. Carcinog. Risks Hum. 90 (2007) 1–636.
[22] I. Frazer, Strategies for immunoprophylaxis and immunotherapy of papillomaviruses, Clin. Dermatol. 15 (2) (1997) 285–297.
[23] M.R. Hilleman, Overview of viruses, cancer, and vaccines in concept and in reality, Recent Results Cancer Res. 154 (1998) 345–362.
[24] S.L. Katz, Future vaccines and a global perspective, Lancet 350 (9093) (1997) 1767–1770.
[25] R. Herrero, et al., Human papillomavirus and oral cancer: the International Agency for Research on Cancer multicenter study, J. Natl. Cancer Inst. 95 (23) (2003) 1772–1783.
[26] A.R. Kreimer, et al., Human papillomavirus types in head and neck squamous cell carcinomas worldwide: a systematic review, Cancer Epidemiol. Biomarkers Prev. 14 (2) (2005) 467–475.
[27] G.M. Clifford, et al., Human papillomavirus genotype distribution in low-grade cervical lesions: comparison by geographic region and with cervical cancer, Cancer Epidemiol. Biomarkers Prev. 14 (5) (2005) 1157–1164.
[28] R.M. Zinkernagel, On natural and artificial vaccinations, Annu. Rev. Immunol. 21 (2003) 515–546.
[29] F.X. Bosch, et al., The causal relation between human papillomavirus and cervical cancer, J. Clin. Pathol. 55 (4) (2002) 244–265.
[30] F.X. Bosch, et al., Prevalence of human papillomavirus in cervical cancer: a worldwide perspective. International biological study on cervical cancer (IBSCC) study group, J. Natl. Cancer Inst. 87 (11) (1995) 796–802.
[31] N.R. Chu, Therapeutic vaccination for the treatment of mucosotropic human papillomavirus-associated disease, Expert Opin. Biol. Ther. 3 (3) (2003) 477–486.
[32] M.A. Stanley, Genital human papillomavirus infections: current and prospective therapies, J. Gen. Virol. 93 (Pt 4) (2012) 681–691.
[33] M. Ling, et al., Preventive and therapeutic vaccines for human papillomavirus-associated cervical cancers, J. Biomed. Sci. 7 (5) (2000) 341–356.
[34] J. Maclean, E.P. Rybicki, A.L. Williamson, Vaccination strategies for the prevention of cervical cancer, Expert Rev. Anticancer Ther. 5 (1) (2005) 97–107.

[35] L.A. Pinto, et al., HPV-16 L1 VLP vaccine elicits a broad-spectrum of cytokine responses in whole blood, Vaccine 23 (27) (2005) 3555–3564.

[36] J.T. Schiller, D.R. Lowy, Papillomavirus-like particles and HPV vaccine development, Semin. Cancer Biol. 7 (6) (1996) 373–382.

[37] W. Ma, C.J. Melief, S.H. van der Burg, Control of immune escaped human papilloma virus is regained after therapeutic vaccination, Curr. Opin. Virol. 23 (2017) 16–22.

[38] C.J. Melief, et al., Therapeutic cancer vaccines, J. Clin. Invest. 125 (9) (2015) 3401–3412.

[39] R.P. Araldi, et al., Papillomaviruses: a systematic review, Genet. Mol. Biol. 40 (1) (2017) 1–21.

[40] K.A. Aires, et al., Production of human papillomavirus type 16 L1 virus-like particles by recombinant *Lactobacillus casei* cells, Appl. Environ. Microbiol. 72 (1) (2006) 745–752.

[41] K. Dell, et al., Intranasal immunization with human papillomavirus type 16 capsomeres in the presence of non-toxic cholera toxin-based adjuvants elicits increased vaginal immunoglobulin levels, Vaccine 24 (13) (2006) 2238–2247.

[42] B. Johns, T.T. Torres, C. Who, Costs of scaling up health interventions: a systematic review, Health Policy Plan. 20 (1) (2005) 1–13.

[43] P. Musgrove, J. Fox-Rushby, Cost-effectiveness analysis for priority setting, in: D.T. Jamison, J.G. Breman, A.R. Measham (Eds.), Disease Control Priorities in Developing Countries, Oxford University Press, Washington, DC, 2006.

[44] R.C. Rose, et al., Oral vaccination of mice with human papillomavirus virus-like particles induces systemic virus-neutralizing antibodies, Vaccine 17 (17) (1999) 2129–2135.

[45] D. Fraillery, et al., *Salmonella enterica* serovar Typhi Ty21a expressing human papillomavirus type 16 L1 as a potential live vaccine against cervical cancer and typhoid fever, Clin. Vaccine Immunol. 14 (10) (2007) 1285–1295.

[46] B. Karanam, et al., Developing vaccines against minor capsid antigen L2 to prevent papillomavirus infection, Immunol. Cell Biol. 87 (4) (2009) 287–299.

[47] M. Tyler, et al., Immunization with a consensus epitope from human papillomavirus L2 induces antibodies that are broadly neutralizing, Vaccine 32 (34) (2014) 4267–4274.

[48] M. Boxus, et al., Broad cross-protection is induced in preclinical models by a human papillomavirus vaccine composed of L1/L2 chimeric virus-like particles, J. Virol. 90 (14) (2016) 6314–6325.

[49] K. Adachi, et al., Oral immunization with a *Lactobacillus casei* vaccine expressing human papillomavirus (HPV) type 16 E7 is an effective strategy to induce mucosal cytotoxic lymphocytes against HPV16 E7, Vaccine 28 (16) (2010) 2810–2817.

[50] L. Cory, C. Chu, ADXS-HPV: a therapeutic Listeria vaccination targeting cervical cancers expressing the HPV E7 antigen, Hum. Vaccin. Immunother. 10 (11) (2014) 3190–3195.

[51] B. del Rio, et al., Oral immunization with recombinant *Lactobacillus plantarum* induces a protective immune response in mice with Lyme disease, Clin. Vaccine Immunol. 15 (9) (2008) 1429–1435.

[52] J.L. Brun, et al., Regression of high-grade cervical intraepithelial neoplasia with TG4001 targeted immunotherapy, Am. J. Obstet. Gynecol. 204 (2) (2011). 169.e1-8.

[53] S. Kumar, M. Biswas, T. Jose, HPV vaccine: current status and future directions, Med. J. Armed Forces India 71 (2) (2015) 171–177.

[54] P. Jin, E. Wang, Polymorphism in clinical immunology—from HLA typing to immunogenetic profiling, J. Transl. Med. 1 (1) (2003) 8.

[55] G.G. Kenter, et al., Vaccination against HPV-16 oncoproteins for vulvar intraepithelial neoplasia, N. Engl. J. Med. 361 (19) (2009) 1838–1847.

[56] A. Yang, et al., The current state of therapeutic and T cell-based vaccines against human papillomaviruses, Virus Res. 231 (2017) 148–165.

[57] W.W. Leitner, H. Ying, N.P. Restifo, DNA and RNA-based vaccines: principles, progress and prospects, Vaccine 18 (9–10) (1999) 765–777.

[58] K. Lin, et al., Perspectives for preventive and therapeutic HPV vaccines, J. Formos. Med. Assoc. 109 (1) (2010) 4–24.
[59] M. Matijevic, et al., Immunization with a poly (lactide co-glycolide) encapsulated plasmid DNA expressing antigenic regions of HPV 16 and 18 results in an increase in the precursor frequency of T cells that respond to epitopes from HPV 16, 18, 6 and 11, Cell. Immunol. 270 (1) (2011) 62–69.
[60] O. Haigh, J. Kattenbelt, M. Cochrane, et al., Hepatitis B surface antigen fusions delivered by DNA vaccination elicit CTL responses to human papillomavirus oncoproteins associated with tumor protection, Cancer Gene Ther. 17 (2010) 708–720.
[61] L. Zhou, et al., Long-term protection against human papillomavirus e7-positive tumor by a single vaccination of adeno-associated virus vectors encoding a fusion protein of inactivated e7 of human papillomavirus 16/18 and heat shock protein 70, Hum. Gene Ther. 21 (1) (2010) 109–119.
[62] P. Berglund, et al., Outcome of immunization of cynomolgus monkeys with recombinant Semliki Forest virus encoding human immunodeficiency virus type 1 envelope protein and challenge with a high dose of SHIV-4 virus, AIDS Res. Hum. Retroviruses 13 (17) (1997) 1487–1495.
[63] J.L. Brandsma, et al., Vesicular stomatitis virus-based therapeutic vaccination targeted to the E1, E2, E6, and E7 proteins of cottontail rabbit papillomavirus, J. Virol. 81 (11) (2007) 5749–5758.
[64] M.J. Hariharan, et al., DNA immunization against herpes simplex virus: enhanced efficacy using a Sindbis virus-based vector, J. Virol. 72 (2) (1998) 950–958.
[65] P. Pushko, et al., Replicon-helper systems from attenuated Venezuelan equine encephalitis virus: expression of heterologous genes in vitro and immunization against heterologous pathogens in vivo, Virology 239 (2) (1997) 389–401.
[66] K.F. Hsu, et al., Enhancement of suicidal DNA vaccine potency by linking *Mycobacterium tuberculosis* heat shock protein 70 to an antigen, Gene Ther. 8 (5) (2001) 376–383.
[67] T.W. Kim, et al., Enhancement of suicidal DNA vaccine potency by delaying suicidal DNA-induced cell death, Gene Ther. 11 (3) (2004) 336–342.
[68] W.F. Cheng, et al., Enhancement of Sindbis virus self-replicating RNA vaccine potency by linkage of *Mycobacterium tuberculosis* heat shock protein 70 gene to an antigen gene, J. Immunol. 166 (10) (2001) 6218–6226.
[69] B.W. Tillman, et al., Adenoviral vectors targeted to CD40 enhance the efficacy of dendritic cell-based vaccination against human papillomavirus 16-induced tumor cells in a murine model, Cancer Res. 60 (19) (2000) 5456–5463.
[70] R. Mikyskova, et al., Treatment of minimal residual disease after surgery or chemotherapy in mice carrying HPV16-associated tumours: cytokine and gene therapy with IL-2 and GM-CSF, Int. J. Oncol. 24 (1) (2004) 161–167.
[71] S. Peng, et al., Vaccination with dendritic cells transfected with BAK and BAX siRNA enhances antigen-specific immune responses by prolonging dendritic cell life, Hum. Gene Ther. 16 (5) (2005) 584–593.
[72] A.D. Santin, et al., HPV16/18 E7-pulsed dendritic cell vaccination in cervical cancer patients with recurrent disease refractory to standard treatment modalities, Gynecol. Oncol. 100 (3) (2006) 469–478.
[73] J. Bubenik, et al., Interleukin 2 gene therapy of residual disease in mice carrying tumours induced by HPV 16, Int. J. Oncol. 14 (3) (1999) 593–597.
[74] S. Hallez, et al., Interleukin-12-secreting human papillomavirus type 16-transformed cells provide a potent cancer vaccine that generates E7-directed immunity, Int. J. Cancer 81 (3) (1999) 428–437.
[75] E.Y. Chang, et al., Antigen-specific cancer immunotherapy using a GM-CSF secreting allogeneic tumor cell-based vaccine, Int. J. Cancer 86 (5) (2000) 725–730.
[76] L. Backert, O. Kohlbacher, Immunoinformatics and epitope prediction in the age of genomic medicine, Genome Med. 7 (2015) 119.

CHAPTER 10

Available immunotherapies and future opportunities to prevent HPV-associated cancers

Rohit Kumar Tiwari[a], Afza Ahmad[a], Irfan Ahmad Ansari[a], and Preeti Bajpai[b]
[a]Department of Biosciences, Integral University, Lucknow, Uttar Pradesh, India
[b]Department of Zoology, Mahatma Gandhi Central University, Motihari, Bihar, India

1. Introduction

Several forms of malignancies including cervix, vagina, vulva, head and neck, and penile are commonly associated with human papillomavirus (HPV) [1]. More than 100 HPV variants have been identified to date [2]. Of these, the most frequently encountered high-risk HPV (HR-HPV) variants (16, 18, 31, and 45) are together responsible for approximately 80% of cervical cancer (CC) cases [3–5]. Among several other variants, HPV-16 and -18 are considered as most pervasive HPV variants that are linked with an estimated 62.6% and 15.7% of several malignancies, respectively [6]. Intriguingly, these stated variants are responsible for $<85\%$ of vaginal and vulvar, $\sim 90\%$ of oropharyngeal, 93% of anal as well as $\sim 80\%$ of penile carcinomas [7]. HPV infection in women accounts for an alarming situation globally as its infection increases the risk of developing cancer [8], thus making HPV a current public health priority (Fig. 1). Approximately 291 million HPV-positive incidences were reported globally in 2007 [9]. In the women exhibiting homeostatic cytology of the cervix, the prevalence of HPV infection was 10.4% and 11.7% between 2007 and 2010, respectively. Ethnically, increased reports of HPV infection were reported in women belonging to Oceania (30.9%), Africa (slightly more than 20%), persuaded by Europe (14.2%), America (11.5%), and Asia (9.4%) [10,11]. Moreover, among the female population globally, 32.1% were reported to be the carriers of HPV infection in 2011 whereas the females belonging to Asia and Africa demonstrated higher prevalence of 45.5% and 29.6%, respectively [12].

Immunopathology, Diagnosis, and Treatment of HPV-induced Malignancies
https://doi.org/10.1016/B978-0-323-91797-1.00001-6

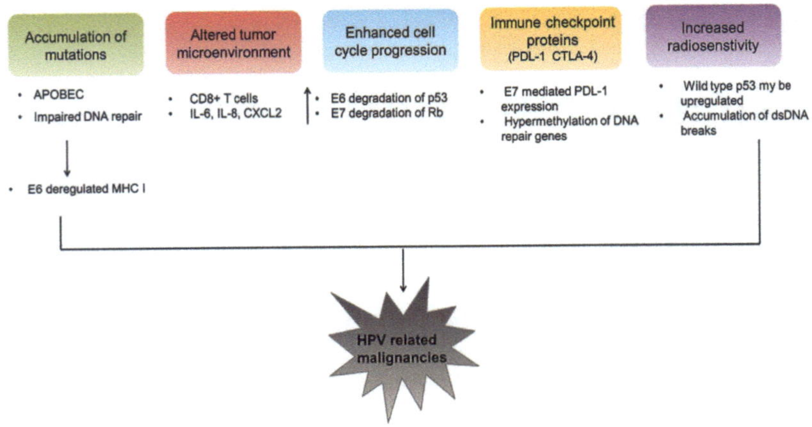

Fig. 1 Schematic representation of specific attributes common within various HPV-related malignancies

Nearly, half of the HPV prevalence as compared with the global prevalence is reported from Asian countries whereas the least number of HPV carriers were reported from Europe (<30%) with the least being from the Western Europe (3.7%). These incidences have clearly outlined that the HPV infection is at surge in the developing nations (42.2%) in juxtaposition with developed nations of the world (22.6%) [13–15]. Additionally, all these studies have shown similar trends in terms of age. Females falling under the age group of 10–25 years are reported to be highly susceptible for HPV infections [12,16]. Intriguingly, rebound in HPV infection has also been reported in adults above 45 years in the African (East and West Africa) and American (Central and Southern America) regions [17].

Worldwide prevalence of HPV infection within male genitals is nearly overlapping to that of females along with the transmission rate of the virus [18,19]. The similarity between the prevalence of HPV infection is attributed to primarily the transmission of virus during sexual activities [19]. A study performed in 2014 outlined that at least 9% of 4,065 healthy males with different ethnicity served to be a carrier of HPV [20]. Sexual orientation of males is also accredited with escalating the prevalence of HPV. Homosexual, HPV^+ males fall in high risk category of HPV infection with 90% incidence relatively with the heterosexual males where the chances of developing the infection are influenced by the number of sexual partners. Furthermore, the rate of HPV infection is quite similar in individuals belonging to different age groups [20,21]. Geographically, the occurrence of HPV infection within men is in close similarity with females since high incidences are documented in African male annually (17.2%) which relatively is lesser in annual cases of HPV infection among Asian males (3.2%) [22]. Study of Giuliano and group explicitly showed that increased prevalence of nearly all the HPV genotypes were evident within individuals from the underdeveloped or developing economies contrastingly with their developed counterparts [23]. Reports have further postulated that by 2025, 5% increase in healthy HPV^+ women is globally due to ambiguities persistent in present vaccination and screening programs particularly in underdeveloped and developing economies of the world. Also it should be understood that in spite of men having reduced HPV burden relatively to women, it should not be misinterpreted as men are inferior carrier of HPV.

Cancer immunotherapy is regarded as the management of various disorders through the use of substances that can modulate the immune system of individuals. The advancements and a detailed understanding of cancer-mediated immunology in past few decades have substantiated that use of immunotherapy in management of human malignancies could be an effective approach. The impelling of tumor-specific immune mediators has far reaching therapeutical intervention than surgical resection and also effectively mediates immune responsiveness against metastasis. Immunotherapy preferentially does not target the diving tumor cells contrasting with the mechanism of chemo- and/or radio-therapeutics. Among several other attributes, many researchers believe that the quiescent cell division resembles cancer stem cells and thus serves to be more efficient target of immunotherapeutic intervention [24]. Irrespective of its approach, immunotherapeutic interventions appear to be competent in negotiating the frequent debilitating side effects of chemo- and/or radio-therapeutics on homeostatic cells/tissues. The present-day immunotherapeutics are far more specific in their mechanism which includes the use of tumor-specific antigen for skewing the immune response. It relies on the use of specific receptors to neutralize the immune evasion strategies. Recently, several of these therapies have been demonstrated for substantial clinical interest even in patients with advanced stages of malignancies (Fig. 2). The present book chapter highlights the available immunotherapeutic approaches that are presently being explored for efficient clinical management of HPV-related malignancies.

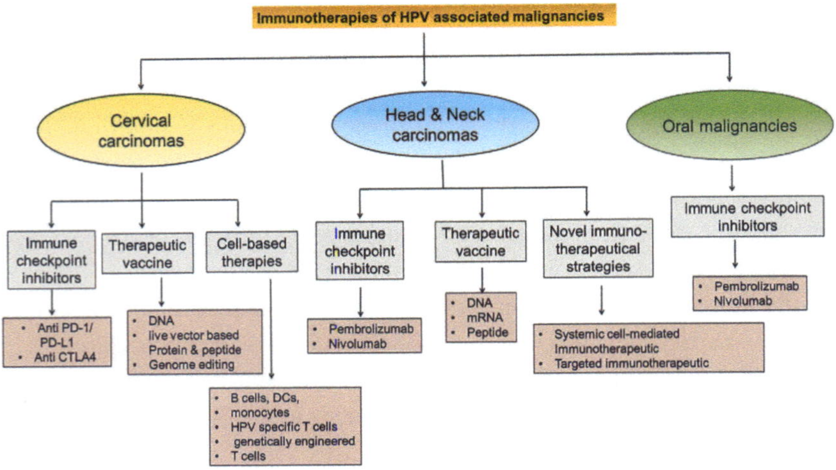

Fig. 2 Flowchart depicting different immunotherapeutic strategies for efficient management of HPV-related malignancies.

2. Present therapeutic options for HPV-associated cancers

Nonmetastatic tumors of genital tract include cervical-, vulvar-intraepithelial neoplasia (C/VIN) and vaginal neoplasia. Presently, the treatment strategy for CIN is primarily constituted by least-invasive surgical excision or cryotherapy [25]. The objective behind such an intervention is the removal of HPV-infected precancerous cells with least stressful impact on cervical integrity and fertility [26]. It is estimated that approximately 528,000 new cases of cervical malignancies are reported annually among which 95% are associated with HPV infections. Similarly, 41% of total 49,000 new cases of vulvar and vaginal malignancies were also linked with HPV infection [27]. Early detection of CC increases the chances of survival for patients, and treatment of such patients relies preferably upon radical hysterectomy or selectively on radical trachelectomy and ionization therapies. Advanced local uncontrolled growth is managed therapeutically through combination of both the chemo- and the radio-therapeutics where the earlier is primarily constituted by cisplatin or its derivatives [28]. However, patients with metastatic cervical malignancies have low survival rates due to compromised renal functioning. For these patients, the standard of care consists of platinum-based chemotherapy doublets, such as cisplatin and paclitaxel [29,30], usually administered in coalition with the humanized monoclonal antibody directed against vascular endothelial growth factor (bevacizumab) [31]. In case of vulvar malignancies, the front-line treatment regime is constituted by surgical resection for nonmetastatic tumors whereas nodal metastatic tumors of vulvar cancer are treated as per the regimes stated in the advanced stage tumors of cervical malignancies [32,33]. Patients with clinical presentation of vaginal malignancies are subjected to standard treatment regimens of primarily three types which are constituted by laser surgery or localized excision vaginectomy, hysterectomy, radiotherapy either external and/or internal as well as systemic or local chemotherapy [34,35].

Moreover, the therapeutical option for head and neck squamous cell carcinomas (HNSCCs) relies upon not only the malignant stage but is vastly governed by its anatomical site. Nevertheless, the core of the treatment as in several other cancers involves surgical, radiological, and chemotherapeutic-based interventions. In spite of several technological advancements in surgical and radiological paradigm, relapse of treatment is observed in approximately 50% HNSCC patients [36,37]. HNSCC tumors are quite challenging for surgical removal and are highly recurrent or metastatic; therefore, the use of chemotherapeutics is relied upon for their treatment. Chemotherapeutical management of HNSCC patients is often quite poor and has been reported for a low mean survival period of only 6–10 months [38]. Immunotherapeutic interventions targeting checkpoint inhibitors are emerging as a promising strategy for patients with clinical display of recurrent or metastatic HNSCC [39,40]. In spite of these advancements, still the mean survival period of recurrent or metastatic HNSCC patients remains unaltered thereby pressing the need for exploration of immunotherapeutic approaches against HNSCC. Recently in 2019, pembrolizumab got approval from US Food and Drug Administration (FDA) as the front-line therapeutics modulating the expression of programmed death ligand 1 (PD-L1) within tumor immune microenvironment for HNSCC patients [41].

3. Immunotherapies for HPV-associated cancers

The history of cancer-related immunotherapies dates long back and includes the initial use of peptide-based vaccines with adaptive immunotherapies to counter various malignancies; however, only a bleak success ratio was attained using these early approaches. Moreover, patients that are treated with immunotherapy have shown varying response rates among cancers and within cohorts of the same malignancy [42]. Immunotherapy is usually accompanied with nonuniform response within individuals owing to its specificity of evoking desired immune response. Furthermore, variations within impediment of cancer cell immune escape strategy and inequitable access of immune cells activated through different mechanisms to the tumor site is further inevitable. These varying responses against immunotherapeutical interventions can further be improved and are not restricted to delineation of efficient biomarkers and inhibitors on immune checkpoints. Furthermore, utilization of predictive tools could certainly increase the response toward these therapies by identifying the plausible responses of patients [43]. Presently, cancer-targeting immunotherapy is being widely explored globally along with supplemented chemotherapy. The reason behind this surge in exploring various immunotherapies against cancer is because these act intrinsically by sensitizing the immune functioning of the affected individuals and thus evade several side effects of chemo- and radio-therapeutics.

4. CC

Among all the clinical malignancies within women, cancers of uterus cervix are the fourth most frequent type. During 2020, 604,127 or 3.1% of total 19,292,789 reported cancers were of cervical and 341,831 total casualties were reported from the same malignancy [44]. Standard treatment option frequently used for clinical management of CC includes surgical removal of tumor growth; however, patients at advanced stages of cervical malignancy also follow the chemo- and/or radio-therapeutical regimes but have exhibited very limited success rate [45]. However, supplementation of platinum-based chemotherapeutics was reported to enhance the survival of patients by 4 months during gynecologic oncology group (GOG)-204 study [31]. Intriguingly, in past couple of decades, technological advancements have led to a better understanding of cell-mediated immune response during HPV infection, which has subsequently led to exploration of novel therapeutical candidates [46,47].

4.1 Immune checkpoints inhibitors

Immune checkpoint inhibitors targeting two negative immune regulatory pathways of T cells such as the programmed death receptor 1 (PD-1) and its ligand PD-L1 and the cytotoxic T-lymphocyte-associated protein 4 (CTLA-4) have been studied in clinical trials as having a high therapeutic index. Furthermore, numerous clinical trials are going on with anti-PD-1/-PDL-1 and anti-CTLA-4. These checkpoint inhibitors are used individually as well as in combination with chemotherapy, chemoradiation, and antiangiogenic agents. Various ongoing clinical trials associated with anti-PD-1/ -PDL-1 and anti-CTLA-4 for HPV^+ cancers are represented in Table 1, respectively.

4.1.1 Anti-PD1/PD-L1 monoclonal antibodies

Nivolumab (fully human-based antibody against PD-1) has been studied as a monotherapy in patients suffering from recurrent or metastatic CC (NCT02257528). The findings of this study demonstrated that individual administration of nivolumab was safe but found associated with low antitumor effect [48]. However, the preliminary findings of another phase I/II clinical study (NCT02488759) have demonstrated that the efficacy of nivolumab monotherapy is propitious in metastatic vaginal, vulvar, and cervical carcinomas [49]. The efficacy and safety of nivolumab along with chemoradiation (NCT03298893) and in combination with paclitaxel (NCT04282109) are being evaluated in patients suffering from locally advanced CC or recurrent HNSCC, respectively. The outcomes of these reports have not yet been reported, but the combination therapies are anticipated to be more effective than nivolumab individually. Pembrolizumab (humanized antibody targeting PD-1) has demonstrated beneficial outcomes in a clinical phase II study as monotherapy of metastatic solid tumors involving cervical and vulvar cancer. The anticancer activity of pembrolizumab in patients with PD-L1-positive advanced CC was steady, and the toxicity was amenable (NCT02628067) [50]. Consequently, these findings expedited the approval of pembrolizumab from FDA for the treatment of PD-L1-positive advanced CC suffering. Furthermore, pembrolizumab monotherapy exhibited promising outcomes in phase III human trials in PD-L1-positive recurrent or metastatic HNSCC patients. The published literature also supports the use of pembrolizumab in combination with 5-fluorouracil and platinum chemotherapy as a front-line treatment for HNSCC [51]. Moreover, another phase II report demonstrated the efficacy of durvalumab (anti-PD-L1) in combination with a personalized cellular immunotherapy known as vigil in advanced cancers involving CC (NCT02725489). Recent reports reveal that vigil comprises of irradiated autologous cancer cells engineered genetically to inhibit the expression of transforming growth factors such as TGFβ1 and TGFβ2 along with recombinant human granulocyte-macrophage colony-stimulating factor (rhGM-CSF). Another combinatorial therapy involving chemotherapeutic drugs such as carboplatin, cyclophosphamide, and atezolizumab (anti-PD-L1) is also in phase I human trial (NCT02914470). Furthermore, atezolizumab has also proven its efficacy in combination

Table 1 Current clinical trials underway for various immunotherapeutical interventions focusing on immune checkpoint inhibitors against HPV-associated malignancies.

Conditions	Agent	Type	Clinical phase	Clinical trial reference
Recurrent or metastatic CC	Nivolumab	Anti PD-1	II	NCT02257528
Various advanced cancer	Nivolumab/ ipilimumab/ daratumumab/ relatlimab	Anti-PD-1/ Anti-CTLA-4/ Anti-CD 38/ Anti-LAG-3	I/II	NCT02488759
CC, locally advanced CC	Nivolumab/ cisplatin/ radiotherapy	Anti-PD-1/ chemotherapy	I/II	NCT03298893
Recurrent HNSCC, head and neck cancer stage IV	Nivolumab/ cisplatin/ radiotherapy	Anti-PD-1/ chemotherapy	II	NCT04282109
CC	Pembrolizumab / platinum / brachytherapy	Anti-PD-1 / chemotherapy / radiotherapy	I	NCT03144466
CC	Pembrolizumab / cisplatin / brachytherapy	Anti-PD-1 / chemotherapy / chemo radiation	II	NCT02635360
CC	Pembrolizumab / paclitaxel cisplatin carboplatin / bevacizumab	Anti-PD-1/ platinum-based chemotherapy / anti-angiogenic agents	II	NCT03367871
CC, endometrial cancer, uterine cancer	Pembrolizumab / immunomodulatory cocktail/radiotherapy	Anti-PD-1 / vitamin D, aspirin, Cyclophosphamide, and lansoprazole	II	NCT03192059
Advanced cancer, CC	AGEN2034 (RebmAb-700)	Anti-PD1	I/II	NCT03104699

SCC recurrent or metastatic, platinum-refractory CC	REGN2810 (cemiplimab)	Anti-PD-1	I/II	NCT03257267
Breast cancer, ovarian cancer fallopian tube cancer	Durvalumab / vigil	Anti-PD-L1 / cellular immunotherapy	II	NCT02725489
HPV, CC, OPC	M7824	Bifunctional antibody targeting TGF-β and PD-L1	II	NCT03427411
Breast cancer, CC, ovarian cancer, endometrial cancer	Atezolizumab/ carboplatin/ cyclophosphamide	Anti-PD-L1/ chemotherapy	I	NCT02914470
Cervical adenocarcinoma, cervical squamous cell carcinoma	Atezolizumab / bevacizumab	Anti-PD-L1 / anti-angiogenic agents	II	NCT02921269
CC, Stage IVB	Atezolizumab / cisplatin/ paclitaxel / bevacizumab	Anti-PD-L1 / chemotherapy / anti-angiogenic agents	III	NCT03556839
Cervical adenocarcinoma, cervical squamous cell carcinoma	Ipilimumab	Anti-CTLA-4	II	NCT01693783
Cervical adenocarcinoma cervical squamous cell carcinoma,	Ipilimumab / cisplatin / radiation	Anti-CTLA-4/ chemotherapy	I	NCT01711515
HPV-mediated (p16-positive) oropharyngeal carcinoma AJCC v8	Nivolumab / ipilimumab / radiotherapy	Anti-PD-1 / anti-CTLA-4 / radiotherapy	II	NCT03799445
Advanced cervical adenocarcinoma advanced vaginal/vulvar carcinoma	Anti-CTLA-4/ anti-PD-L1 / radiotherapy	Anti-CTLA-4/ anti-PD-L1 / radiotherapy	I	NCT03452332

with cisplatin/paclitaxel/bevacizumab (standard therapeutic agents for stage IVB cervical carcinoma) and has entered the phase III clinical trial (NCT03556839) [52].

4.1.2 Anti-CTLA-4 monoclonal antibodies

The CTLA-4 downregulates the cytotoxic responses against tumor cells upon binding to B7-1/B7-2 proteins on antigen-presenting cells. Recently, it has been demonstrated that single-nucleotide polymorphisms (SNPs) in the promoter region of the CTLA-4 gene correlate with greater susceptibility toward different types of carcinomas including CC [53]. Blocking CTLA-4 with an immune checkpoint inhibitor will re-instigate the cytotoxic T-cell population toward killing of tumor cells. Recently developed ipilimumab is a human-based anti-CTLA-4 monoclonal antibody effective against the metastatic or persistent CC. The antibody has entered into different phases of clinical trials (NCT01693783). Preliminary findings on 34 patients exhibited 3 partial responses and 8 cases with a steady disease, whereas 23 patients revealed disease progression. Indeed, ipilimumab was highly compatible in patients with CC and instigated strong immune response [54]. Another clinical phase I study demonstrated the efficacy of cisplatin followed by ipilimumab along with chemoradiation in patients suffering from advanced CC. The results demonstrated high efficiency and compatibility of ipilimumab [55]. Furthermore, another approach involves the utilization of anti-PD-1/PD-L1 and anti-CTLA-4 antibodies. In this regard, an active clinical phase I trial was planned to assess the safety and tolerance of durvalumab and tremelimumab (anti-CTLA-4 antibody) combination in numerous advanced solid tumors such as CC (NCT01975831). Initial findings revealed the promising efficacy and safety of this combination [56]. Another combinatorial therapy involving three monoclonal antibodies, viz. durvalumab, tremelimumab, and vinorelbine (agents which suppress mitosis by interacting with tubulin protein) is currently under clinical evaluation for its use against advanced solid tumors (NCT03518606). Moreover, a combination of duo tremelimumab and durvalumab along with radiotherapy has been analyzed for the management of cervical, vaginal, or vulvar carcinomas (NCT03452332).

4.2 Therapeutic vaccines

Majority of HPV vaccines target E6 and E7 which are crucial oncoproteins of HPV16 and HPV18, the high-risk types of cervical, vaginal, and vulvar cancers in women. [57]. A myriad of approaches has been employed to formulate immune therapeutic agents by using advanced genome editing tools and engaging compatible vectors, peptides/proteins, and DNA. Vector, peptides, or protein vaccines target HPV16, whereas DNA vaccines and modern editing tools act as polyvalent vaccines targeting oncoproteins E6 and E7 of both subtypes of HPV such as HPV16 and HPV18. Presently, several HPV-associated oncoproteins are being explored for their prophylactic potential and are summarized in Table 2.

4.2.1 Vector vaccines

There are definite numbers of recombinant prokaryotic vectors which are responsible for antigen presentation, and they undergo replication within the cells [58]. ADXS11-001 is one such bacterial vector of live attenuated *Listeria monocytogenes* (LM) expressing a HPV 16 E7 oncoprotein. This oncoprotein is linked to LM Listeriolysin O and has been assessed in various clinical studies for the management of $HPV16^{+}$ carcinomas including CC. ADXS1-001 monotherapy has demonstrated propitious outcomes in patients having recurrent CC (NCT01266460). The initial level of these two-stage phase II clinical trial exhibited overall 12-month subsistence of 38.5% patients [59]. On the other side, its second stage revealed that 34.9% patients with 12 months consolidated survival rate thereby promoting subsequent clinical investigations of ADXS11-001 [60]. Furthermore, this therapeutic agent was further compared with placebo in phase III clinical trials of advanced cervical carcinoma patients (NCT02853604). ADXS11-001 has also been investigated in HPV^{+} HNSSC to assess its efficacy as immunostimulant before surgery (NCT02002182). Recently, accomplished phase I/II study with ADXS11-001 exhibited 1×10^{9} colony-forming units (CFUs) as safe dose alone or in combination with cisplatin for administration in patients suffering from advanced anal cancer (NCT01671488) [61]. Moreover, another study recommended the use of higher concentration of ADXS11-001 (1×10^{10} CFU) in different stages of CC (NCT02164461). The vaccine was also administered along with durvalumab against advanced or metastatic cervical/head and neck cancer, and phase I/II clinical trial results were highly promising (NCT02291055) [62]. ADXS11-001 has displayed beneficial outcomes in both the forms as alone or in combination, thereby paving a roadway for its approval as an alternative therapeutic agent for HPV^{+} cancer patients.

4.2.2 Protein and peptide vaccines

The peptide vaccines have been designed from HPV antigens as a ligand for major histocompatibility complex class I (MHC-I)-restricted immunogenic

Table 2 Ongoing clinical trials for different prophylactic interventions against HPV-associated malignancies.

Conditions	Agent	HPV target	Type/stage	Clinical phase	Clinical trial reference
HR CC, advanced CC	ADXS11-001/ placebo	HPV 16 E7	Bacterial vector	III	NCT02853604
Effects of immunotherapy, metastatic/ recurrent CC, cervical adenocarcinoma	ADXS11-001	HPV 16 E7	Bacterial vector	I/III	NCT02164461
HPV-related carcinoma HPV-related CC, HPV-related anal squamous cell carcinoma	TG4001 avelumab	HPV-16 E6/E7	Viral vector anti-PD-L1	I/II	NCT03260023
HPV$^+$ vulvar, vaginal, penile, rectal, and anal cancer	PRGN-2009/M782	HPV-16 E6/E7	Viral vector / bifunctional antibody targeting TGF-β and PDL1	I/II	NCT04432597
CC	ISA 101 / carboplatin paclitaxel / bevacizumab	HPV-16 E6/E7	Peptide vaccine / chemotherapy / antiangiogenic	I/II	NCT02128126
Anal cancer, CC, esophageal cancer	HPV16 E7 peptide, synthetic HPV 16 E6 peptide	HPV-16 E6/E7	Peptide vaccine	I	NCT00019110
Squamous cell carcinoma	Paclitaxel poliglumex, cetuximab, radiation therapy (IMRT or 3D-CRT)	–	Stage III–IV LA-HNSCC	I/II	NCT00660218

HNSCC,	Cetuximab, laboratory biomarker analysis, tivantinib	–	R/M HNSCC, first line	II	NCT01696955
HNSCC	Palbociclib, cetuximab, placebo	HPV(–)	Cetuximab-naïve R/M HNSCC, second line	II	NCT02499120
Melanoma	Enoblituzumab, ipilimumab	–	Advanced B7-H3-expressing HNSCC	I	NCT02381314
HNSCC	Cetuximab + IMRT + ipilimumab	–	Stages III–IV HNSCC p16 or intermediate risk p16+	I	NCT01860430
HNSCC	Docetaxel and pembrolizumab	–	LA unresectable or R/M platinum refractory HNSCC	I	NCT02718820
HNSCC	Neoadjuvant pembrolizumab and adjuvant RT ± platinum + pembrolizumab in high-risk patients	–	Stage III/IV resectable HNSCC	II	NCT02296684
HNSCC	Pembrolizumab, Intensity-modulated radiation therapy	–	Stages III–IV (nonmetastatic) HNSCC	II	NCT02609503

epitopes. One of them is DPX-E7 (9-amino-acid-long synthetic peptide) consisting of HPV 16 E711-19 and packaged into liposomes and stored in freeze-dried condition. The efficacy of DPX-E7 has been investigated in an open-label phase I/II human trials involving HLA-A*02-01 patients with HPV16- associated anal, head and neck, and CC (NCT02865135). Another synthetic peptide ISA 101 comprising of 12 surface layer proteins (SLPs) having 25–35 residues obtained from E6 and E7 oncoproteins of HPV16 genome has been evaluated alone and along with nivolumab in patients suffering from HPV16$^+$ solid tumors (NCT02426892). As demonstrated by the ORR and OS, the promising results have been obtained specifically in comparison with the individual administration of anti-PD1 [63]. Moreover, another clinical study exhibited the employment of ISA 101 along with pegylated interferon (IFN-α IIb), and the findings revealed a noteworthy correlation between OS and T-cell responses triggered by the combination [64]. Furthermore, a vaccine named Hespecta comprising of Toll-like receptor (TLR)-2 ligand conjugated to the HPV16 E6-derived SLPs has also been well studied and progressed for evaluation through phase I clinical trial (NCT02821494).

4.2.3 DNA vaccines

The new-generation DNA vaccines have been widely investigated in various clinical studies and demonstrated promising results; however, their weak immunogenicity is a matter of concern [65,66]. In this direction, numerous approaches have been designed to advance the presentation of DNA vaccine-encoded antigens. GX-188E is one such immunotherapeutic agent which is comprised of tissue plasminogen activator signal sequence, an FMS-like tyrosine kinase 3 ligand and shuffled E6 and E7 genes of HPV16 and HPV18, respectively. GX-118E is administered through intramuscular route by electroporation and has been extensively studied in various trials against cervical intraepithelial neoplasia (CIN) II/III (NCT02139267, NCT02596243). A recent report also indicates its efficiency of this agent in a cohort of 64 CIN III patients [67]. In addition, an interventional study stating the competence of GX-188E in combination with GX-I7, interleukin 7 (IL-7), or imiquimod has also been reported (NCT03206138).

4.2.4 Genome editing tools

Numerous upcoming genome editing tools have revolutionized the research and development of novel agents to combat cancer. Some editing tools, viz. CRISPR/Cas9, zinc finger nucleases (ZFNs), and transcription activator-like effector nuclease (TALENs) have demonstrated promising results by significantly reducing carcinogenesis in HPV16/18 in vitro and in vivo models [68–70]. Accordingly designed chemotherapeutic agents named ZFN-603 and ZFN-758 have been investigated in a phase I clinical study for the treatment of HPV16 or $HPV18^+$ and have revealed promising results by cleaving HPV16/18 E7 oncogene consequently decreasing the expression of E7 resulting in apoptosis of cancer cells (NCT02800369). Another alternative, TALEN and CRISPR/cas9 targeting HPV16 and HPV18 E6/E7, has also been studied in clinical trials for treatment of HPV^+ CIN patients (NCT03057912).

4.3 Cellular therapies

Cell-based therapy involves the isolation of dendritic cells, B and T lymphocytes from the patient, and thereafter ex vivo transduction to express or target a specific antigen before being reinfused in the patient. Various clinical reports employing adoptive T-cell chemotherapy (ATC) approaches for management of metastatic or recurrent CC have exhibited propitious outcomes. Stevanovic et al. performed the first study which demonstrated the promising results of ATC against CC with tumor-infiltrating lymphocytes (TILs) expanded ex vivo [71]. Another study by Ramos et al. [72] verified that HPV-specific T lymphocytes (HPVST) isolated from patients could be beneficial to combat the disease. The tailored HPVST resistant to TGF-β are currently under phase I human trial (NCT02379520) for their safety efficacy evaluation. Recently, HPVST has also been used in combination with nivolumab for the treatment of recurrent HPV-associated carcinomas. Reports have demonstrated that HPV-TILs could cause tumor suppression in HPV-associated cancers (NCT01585428) [73]. However, this immunotherapeutic strategy exerts a huge economic burden for management of disease due to the involvement of isolation and maintenance of tumor-specific T cells. Hence, a readymade approach with genetically engineered T cells is essentially required [74].

4.3.1 Chimeric antigen receptor (CAR)-T cells

Chimerically tailored T cells expressing a single chain of variable fragment (scFv) from human origin are capable of recognizing a tumor-associated antigens. These cells employ CD3 receptor-associated molecules and co-stimulatory domains as intracellular signaling machinery. The chimeric antigen receptor T cells display antigen binding in an MHC-independent manner [75]. Numerous therapeutic agents have been designed using the aforementioned concept, and few of them have even reached advanced stages of clinical trials. A recent clinical trial (NCT03356795) focused on investigation of safe CAR T cells use in CC patients (cancer cells expressing tumor-associated antigens (TAAs), viz. ganglioside GD2, prostate-specific membrane antigen (PSMA), mucin-1 or mesothelin) has revealed promising results. In this study, the T cells are isolated from peripheral blood mononuclear cells (PBMCs) through apheresis from CC patients positive for the respective TAAs and genetically altered in in vitro conditions to become TAA-specific CAR-T cells. The tailored cells are thereby infused intravenously into the patients. The proliferation, persistence, and activation of CAR-T cells in the peripheral blood have also been evaluated by the research group.

4.3.2 Engineered T-cell receptor therapy

The most appropriate oncoproteins of HR-HPVs are nuclear and are absent on the surface of cancer cells. Even though they include epitopes being processed and presented by MHC and recognized by T-cell receptors (TCRs), recognition of epitopes by engineered TCR is advantageous over the use of CAR T cells [76]. Numerous approaches have been delineated to determine tumor-specific T cells, viz. (i) isolation of HPV reactive T cells from patients suffering from tumor regression, (ii) immunization with human oncogenic proteins obtained from transgenic murine models expressing human HLA, and (iii) generation of cytotoxic T lymphocytes in in vitro conditions [77–79].

The isolation of T-cell population reactive toward the target pMHC and their respective TCR recognition is carried out by single-cell sequencing. Various techniques such as multiplex PCR, target enrichment, and 5′ rapid amplification of cDNA ends (RACE) based on cDNA library preparation provide the complete transcripts of the TCR chains. Moreover, single-cell sequencing technologies involving single-cell RT-PCR and paired SEQ could determine the TCR population at single-cell level [80]. The process is followed by the identification and cloning of HPV-specific TCRs (TCRα and TCRβ genes) for subsequent characterization. The RNA or plasmid DNA transfection methods are used for gene delivery [81–83]. Owing to its high efficiency and long-term expression, virus-mediated delivery is generally preferred specifically by employing adeno-associated viruses (AAVs), gamma-retrovirus, and lentivirus [84].

5. HNSCC

The sixth most frequent carcinoma accounting for 905,000 new incidences and 385,000 annual deaths globally is HNSCC. The affected body part mainly constitutes of the oropharynx, oral cavity, hypopharynx, and larynx. Long-term consumption of alcohol and tobacco along with the infection of high-risk HPV is regarded as the chief oncogenic drivers [85]. HNSSC is correlated with the low survival rate as majority of the patients are diagnosed with advanced clinical stages when there is a high risk of recurrence and metastasis [86].

The standard therapeutical regimen involving platinum-based chemotherapy, fluorouracil, and cetuximab is used for the treatment of recurrent and/or metastatic (R/M) HNSCC patients who are not susceptible for salvage surgery or radiotherapy [87]. In case of progressive disease, second-line therapeutic options are limited to single-agent chemotherapy including methotrexate, docetaxel, or cetuximab. In spite of the numerous multimodalities employed, the long-term control of the disease is challenging as the response rate (RR) in the second-line treatment is just 4% and in median overall survival (OS) of > 6 months [88]. Among all the available therapeutic options, immunotherapy is a propitious approach which restores the immunosuppressive microenvironment in HNSCC [89]. Antitumor immune alterations in HNSCC comprise reduced count and activity of natural killer cells, followed by poor antigen presentation. Consequently, the dysfunctional T cells express checkpoint molecules, viz. PD-1 and/or CTLA-4, thereby combating the disease [90].

5.1 Immune checkpoint inhibitors

The tolerability of nivolumab and pembrolizumab within HNSCC patients undergoing chemotherapeutical interventions was highly witnessed in the last decade. The humanized IgG monoclonal antibody, nivolumab, competently elevated the overall survival of HNSCC patients during immunotherapeutical trials, thereby receiving "category 1" endorsement from the National Cancer Committee Network [39], whereas pembrolizumab was successful in getting acceptance as of "category 2a" based on the findings of nonrandomized trials [91,92]. Intriguingly, during a recent randomized phase III trial, pembrolizumab proved unsuccessful for its efficacy in comparison with chemotherapeutics [93]. Owing to its multifaceted applicability as a biomarker in several clinical settings, PD-L1 and its augmented expression is closely associated with enhanced survival advantages post-nivolumab and -pembrolizumab treatment. Nevertheless, elucidation of other more suited and specific biomarkers till date remains a substantial challenge to further boost patient selection, response prediction, and escalating the efficacy.

5.1.1 Pembrolizumab

Pembrolizumab also referred to as MK-3475 is a well-recognized monoclonal IgG4 humanized antibody targeted against PD-1. The tolerability and efficiency of pembrolizumab were evaluated initially during an open-label phase Ib clinical setting reported as KEYNOTE-012 trial, within cohorts of HNSCC patients suffering from advanced metastatic tumors [40]. The study revealed the competency of the antibody in instigating chronic and clinically sustained anticancer efficacy within PD-L1-positive patients at a dosage of 10mg/kg. Notably, the positivity of PD-L1 was delineated by its expression (≥1%) on tumor as well as immune cell expression. Moreover, further evaluation of drug efficacy was carried out in 132 patients cohort positive for metastatic HNSCC at a dosage of 200 mg/kg. About 57% of the enrolled patients were reported to have undergone couple of therapies prior to enrollment, and the number of PD-L1-positive patients was found to be 78%. The cumulative objective response rate was evaluated to be 20% whereas the overall survival rate was substantiated as 59%. The study indicated that predetermined and less-frequent dosage regimes have several benefits including waste reduction, tolerability, and convenience [91]. Pembrolizumab established itself for tolerability and safety with most of the adverse events (AEs) delineated to fatigue as most prevalent symptoms falling in grade 1/2. Furthermore, immune-associated AEs of grade 3/4 particularly pneumonitis were relatively uncommon and reported to be <4%. Based on stated observations, the FDA approved the human usage of pembrolizumab in 2016 within patients having metastatic or recurrent HNSCC either concomitantly or postplatinum-based chemotherapeutics.

A recent single-arm phase II trial carried out in 171 patients suffering from recurrent or metastatic HNSCC demonstrated comparative efficacy of pembrolizumab against resistant platinum-based chemotherapeutic and cetuximab [92]. During the trial, it was being observed that out of 171 enrollments, 82% were positive for PD-L1 expression whereas HPV infection was confirmed in 22%. The ORR constituted by complete and partial response (C/PR) was 16% of which one displayed PR whereas stable disease (SD) was reported in 19%. Nearly, overlapping results were observed in either HPV or PD-L1 expressing recruits of the study (16% in HPV confirmed and 15% in negative patients, whereas 18% in PD-L1 positive and 12% in negative recruits, respectively). Additionally, the intermediate duration of response in the study was 8 months which intriguingly was slightly higher in HPV-positive recruits in juxtaposition with respect to their negative counterparts. Majorities of AEs reported in the trial were categorically delineated as either grade 1 or 2. However, AEs of grade 3 and higher were also documented in 15% of recruits which further ended in termination of trials within 4% of recruits whereas one patient died during the trials due to complications from pneumonitis. AEs comprised of hyper- and hypothyroidism and pneumonitis within 2%, 16%, and 4% of enrolled individuals were reported recurrently thereby leading to modifications in further trials. Accordingly, the phase III randomized trial was undertaken to comparatively evaluate pembrolizumab with standard chemotherapeutics administered during metastatic HNSCC postrelapse with platinum-type chemotherapeutics (KEYNOTE-040). The interpretations were further reported at European Society of Medical Oncology (ESMO) 2017 Congress [94]. 495 patients were equitably randomized either to pembrolizumab or docetaxel, methotrexate or cetuximab (based on investigator discretion) served as representative of standard chemotherapeutics. Observations from the study explicitly elucidated that pembrolizumab displayed deterioration in efficacy to fulfill primary end point of OS. Nonetheless, analysis of subset within the cohort enumerated several significant betterments closely related with elevated expression of PD-L1. The study further validated the safety and tolerability of pembrolizumab apart from hypothyroidism which was manifested in 13% patients relative to 1% as documented in chemotherapeutic interventions during the study.

5.1.2 Nivolumab

Nivolumab is monoclonal IgG4 antibody against human PD-1 receptors which had earlier been substantiated for its efficacy and tolerability at a lone dose of 10 mg/kg (every 2 week) during couple of initial clinical trials within metastatic tumor patients [95,96]. The observations from these trials instigated final phase of randomized clinical trial of nivolumab in HNSCC patient refractory toward platinum-based chemotherapeutics (CheckMate-141). During this study, 361 volunteers were bifurcated in the ratio of 2:1 subsequently receiving nivolumab (3 mg/kg body weight, once for every 2 week) or chemotherapeutics, namely docetaxel, cetuximab, or methotrexate. Furthermore, 54.6% recruits in the study were reported previously for couple of systemic therapies. Immunohistochemical findings of the study substantiated that PD-L1 expression was recorded as $\geq 1\%$, $\geq 5\%$, or ≥10% in at least 100 cancerous cells. The trial concluded ahead of its completion time after attaining the primary endpoints. Recruits receiving the nivolumab treatment illustrated elevated mean OS which doubled in 1 year subsequently followed by superior ORR. The median survival of patients without any signs of disease progression was overlapping in both the formulated groups. The safety profile of nivolumab was much appropriate than the chemotherapeutics that were concomitantly used in the trials with AEs belonging to grade III. Commonly observed AEs during the trial were fatigue, nausea, and rashes which were reported in 14%, 8.5%, and 7.6%, respectively, whereas the AEs of grade 3 or 4 included fatigue and anemia within 2.1% and 1.3% individuals, respectively. The impact of nivolumab on health-associated quality of life was reported to be of good standards during an exploratory study [97]. In view of these compelling evidences, nivolumab was approved for human use in 2016 by FDA for metastatic HNSCC patients undergoing or after platinum-derived chemotherapeutics.

5.2 Therapeutic vaccines

Delivery of antigenic epitopes leading to the induction of sustainable T-cell response in an immunogenic context remains the prime objective in vaccination strategy. Recently, multiple cancer vaccines have been underlined, and among them, DNA and RNA vaccines encoding selected tumor antigens or synthetic long peptides (SLPs) vaccines co-delivering CD4 and CD8 epitopes are substantially highlighted ones [98]. These vaccines focus on delivering selective antigens without concomitant delivery of backbone-encoded antigens as frequently encountered during pathogen-extracted vectors including bacteria or viruses. The latter often leads in generation of profound CD8-specific epitopes skewing the responsiveness of immune system toward the vector [99]. However, various viral and bacterial vectors-based prophylactic approaches have shown their clinical competence in instigating CD8-dependent responses in dysplastic disease and several malignancies [100,101].

5.2.1 DNA vaccines

DNA-based vaccines represent a simple strategy for direct delivery of plasmid DNA encoding antigens whose expression is achieved through eukaryotic promoter. This process culminates in synthesis of antigen within the body where the transfected DNA backbone is reported to act as an adjuvant [101]. Several innate immune receptors are reported to specifically recognize plasmid DNA and include intracellular TLR-9 along with cytosolic

DNA-dependent activator of IFN-regulatory factors (DAI), cytosolic DNA sensing, cyclic GMP-AMP synthase stimulator of interferon genes (cGAS-STING), among several others [102]. These prophylactic interventions have numerous benefits which include stability, safety, and relatively simple preparation, although reduced immunogenicity of these preparations has earlier been reported initially in clinical settings [103]. This drawback was further rectified by combinatorial approach of DNA vaccines and through in vivo electroporation (EP). EP has initially been reported for its relevance in escalating the bioavailability of transfected DNA which facilitates profound recognition of DNA by sensors of innate immune cells leading generation of desired cell-mediated immune response [104].

DNA vaccines have also been explored against HPV-mediated infections where they were reported to impede binding of HPV-associated E6 and E7 oncogenes with p53 and pRb individually [105,106]. Several DNA vaccines against HPV infections have been explored in clinical settings in CIN. Recently concluded clinical trial of phase IIb involving VGX-3100 DNA vaccine coding the HPV oncogenes E6 and E7 was administered in combination with another IL-12-coding DNA vaccine with EP elucidated promising outcomes in patients tested positive for HPV-16- and HPV-18-related CIN. 50% of females vaccinated through the stated regime elucidated profound generation of T-cell-mediated immune response with concomitant reduction of the lesions [107]. Owing to its initial success on HPV infection, VGX-3100 was subsequently assessed in clinical settings of HNSCC. However, in this trial also, the vaccine was administered in combinatorial approach with IL-12-coding DNA vaccine along with EP (NCT02163057). The findings delineated that HPV-related E6- and E7-specific antibodies were significantly expressed in 80% of recruits whereas escalated virus-specific cell-mediated immune response was also evident in 9 of the total 10 enrolled patients. Additionally in the trial, it was also confirmed through flow cytometric studies that HPV-specific $CD8^+$ cells with simultaneous expression of granzyme/perforin were also present in 87.5% of enrolled patients [108]. Currently, a phase I/II trial is underway to assess the tolerability and anticancer efficiency of MEDI-0457 or MedImmune in a combinatorial regime with durvalumab (monoclonal antibody against PD-L1) in confirmed patients with HPV-related HNSCC (NCT03162224). pNGVL4a-CRT/E7 (detox) is a DNA-based vaccine against HPV16 E7-calrericulin (CRT) illustrated its potency of augmenting MHC-I presentation with simultaneous anti-angiogenic effects [109]. Another preclinical study demonstrated the administration of TC-1 tumor led to instigation of anti-E7 antibodies and T-cell-dependent response [110]. A pilot clinical research using pNGVL4a-CRT/E7 (detox) for therapeutical management of HPV16-related CIN2/3 was undertaken recently [111]. EP was excluded from this exploration; however, one arm studied the particle-mediated epidermal delivery employing a needle-free ND10 delivery system. The results exhibited mild but manageable toxicity principally localized to administration site, but only a little increase in systemic T-cell responses was noted with no increase in regression above the control. Recently, a DNA vaccine called doggybone (db) DNA (dbDNA, Touchlight Genetics) was recently confirmed for its tolerability and immunogenic potential within HPV16-related E6 and E7 oncogenes preclinical model [112]. dbDNA vaccine worked independent of TLR-9-mediated recognition but exerted its prophylactic effects through STING-mediated signaling. Notably, overlapping levels of T- and B-cell-mediated immune responsiveness were evident in case EP aided delivery of dbDNA and standard plasmid DNA vaccine. This unique DNA-based vaccine represents a propitious alternative to plasmid DNA prophylactics for efficient management of HPV-related E6 and E7 antigens.

5.2.2 mRNA vaccines

Surge in exploring the prophylactic potential of mRNA-based vaccine is being witnessed in recent times. They ensure antigen expression through accumulating at high concentration in the cytoplasm. mRNA is a well-known natural ligand for TLR-3, TLR-7/8, and various cytosolic sensors (i.e., RIG-I, MDA5), which induce innate immune responsiveness required for augmenting the efficiency of the vaccines. The delivery of mRNA in vivo is achieved by its complexation with a lipid-based carrier preventing its degradation and efficiently targeting dendritic cells [113]. Recently, a first-in-human personalized mRNA-based vaccine against melanoma patients mutanome was reported [114]. The responsiveness of T cell in opposition to numerous mutanome-derived neoepitopes was evident among all the patients. It was found that during the initiation of vaccination trials, four out of five progressive metastatic patients demonstrated either partial or complete response clinically postvaccination. Intriguingly, one patient was benefited from vaccination subsequently followed by anti-PD1 mAb. Moreover, phase I/II clinical trial utilizing an mRNA vaccine targeting HPV16 E6 and E7 has recently started (NCT03418480). It is proposed that during the study, HPV-positive HNSCC patients will be administered with the vaccine either alone or following a combinatorial regime constituted by anti-CD40 antibodies.

5.2.3 Peptide vaccines

Peptide-based prophylactic interventions are important type of vaccines which are characterized by their safety and hassle-free reproducibility, although the major drawback associated with these remains their associated costs and low levels of immunogenicity. These vaccines are often constituted by CD8 epitopes against a predefined HLA allele or may include long single and/or overlapping sequences of peptides which may be inclusive of CD4- and CD8-specific epitopes. However, the contemporary approaches have overcome the drawbacks frequently associated with the HLA restriction. Adjuvant-based combinatorial prophylactic regimes are also been substantially explored to enhance the protective efficacy of these mRNA vaccines. After its initial approval in both USA and European countries, squalane oil emulsion or MF59 was explored for its potential in preliminary human trials [115]. Several clinical trials have focused upon the use of different surface layer proteins (SLPs) against HPV-specific antigens in combinatorial prophylactic regimes with oil-in-water-based adjuvants for clinical management of both CIN and vulvar neoplasia which has exhibited significant decline in premalignant lesions [116]. Intriguingly, this strategy failed to exert its protective effects against cervical malignancies and was shown to cause subtle changes in responsiveness of T cells culminating in no significant benefits [117]. This incompetent trial necessitates the urgency of urgent exploration of effective adjuvants which can be used in combinatorial approaches to skew the tumor microenvironment [118].

A clinical trial of phase I/II (NCT02865135) aimed at exploring the tolerability and efficiency of short peptide sequence against HPV-related E7 oncoprotein was recently undertaken in combinatorial regime constituted by cyclophosphamide in HLA-A2 patients exhibiting recurrent malignancies of anus, cervical, and oropharynx with the hypothesis that it would aid in depletion of T_{reg} cells [119,120]. In yet another trial in clinical setting, HPV malignancies-associated patients at advanced stages were weekly administered with prophylactic interventions constituted by p16-derived SLP [121]. The study made it evident that tumor suppression in recruited patients was achieved due to HPV-associated transformations postcompletion of treatment with standard chemotherapeutics. Furthermore, vaccine constituted by CD4 and CD8 epitopes was used as emulsion with Montanide ISA-51 VG which is an oil-in-water class of adjuvant. The study revealed the competency of peptide vaccine in instigating both humoral and cellular immune response with significant tolerability. Of the 14 recruited and vaccinated patients, 9 patients exhibited stable disease; however, 5 were reported for disease progression [122]. Subsequently, the currently active trial NCT02526316 is exploring the effects of same vaccine in case of its administration in various routes, namely subcutaneous and intradermal routes. In addition, clinical trials of phase II level are presently underway which aims to evaluate the prophylactic efficacies of SLPs during coadministration with Nivolumab (NCT02426892) and/or Utomilumab (anti-CD137 antibody with immune stimulatory functions; NCT03258008) in patients exhibiting HPV16-related HNSCC and various other HPV-associated malignancies [118]. Phase I trial of HPV16 E6-related unique SLPs (ISA201) in combination with agonist ofTLR-1/2 Amplivant (NCT02821494) was recently concluded which focused around for impeding premalignant lesions and tumor [123]. A short while ago, multivalent vaccine targeted against HLA-A24 restricted peptides from cancer testis antigens (CTAs) including IMP3, LY6K, and CDCA1 which was administered concomitantly with incomplete Freud's adjuvant cleared clinical trials of phase II level in patients with HNSCC malignancies in Japan [124]. The vaccine was successful in enhancing the OS which was further linked to the instigation of peptide-dependent cytotoxic T lymphocyte response. Interestingly, the patients exhibiting responsiveness against all the three antigens used in the trials were reported to have increased OS relatively to those who exhibited responsiveness against distinctive antigens only.

5.3 Novel immunotherapeutic strategies in HNSCC

5.3.1 *Systemic cell-mediated immunotherapy in HNSCC*

Systemic cell-mediated immunotherapeutic interventions although are nonspecific, yet they focus on replacing the immune responsiveness either through a systemic and/or loco-regional anticancer response. As evidently seen in adoptive transfer therapy, there is proliferation with certain positive alterations within the ex vivo immune cells of the patient subsequently followed by their re-infusion. Initially, this strategy was adopted after success within murine models where curtailments within the size of grafted tumors were reported [125]. Patients at stage IV nasopharyngeal malignancies expressing antigens from Epstein-Barr virus (EBV) were clinically recruited to substantiate the efficacy of earlier stated therapy. During the study, it was shown that the activation and subsequent expansion of T cells (EBV-specific) was achieved ex vivo poststimulation of the blood lymphocytes by sensitizing them with transformed autologous EBV-B cells. The therapy elucidated that the disease progression in 6 among 10 recruited patients was significantly controlled whereas subdued inflammatory responses were observed in 2 patients [126]. Subsequently, researchers have further elucidated the plausibility of producing tumor-specific T cells and the tolerability of this strategy within patients diagnosed at higher stages of HNSCC [127,128]. Similarly, therapy constituted by transfected dendritic cells involves autologous transfection of dendritic cells with DNA of tumor cells derived from the patient followed by its re-infusion. The proof was initially validated within HNSCC patients where it was substantiated that transfection of tumor DNA within dendritic cells led to efficient antigen presentation with simultaneous extermination of indication of dendritic cell suppression [129]. An expansion of this approach exploits intra-tumor dendritic cells with simultaneous immunosuppressive chemo- and radio-therapeutics. Recent translation of this strategy within animal models of HNSCC showed prolonged tumor regression, augmented apoptosis related with reduced intra-tumor regulatory T cells [130].

5.3.2 Targeted immunotherapy in HNSCC

Technological advancements have resulted in elucidation of several tumor-specific antigens which in turn are being explored for their therapeutical relevance in HNSCC patients [131]. Among these antigens, a group is referred to as germ cell antigens or cancer testes antigens (CTAs) which are further characterized by their nearly absent expression within homeostatic tissues but are significantly expressed on certain tumors [132]. It was earlier reported that majority of HNSCC (approximately 71%) positively express specific antigens belonging to either melanoma antigen genes or MAGEs especially MAGE-1 and -3 [133,134]. Similarly, NY-ESO-1 gene is overexpressed in certain types of tumors such as HNSCC comparatively with homeostatic tissues [135,136]. Tumor-related antigens from yet another category known as tumor-specific mutated proteins consequently are involved in development of phenotype associated with malignant tumors such as p53 [131]. Preclinical findings have indicated that mutated p53 is expressed substantially in HNSCC patients and the expression of its wild type predominantly relates with HPV 16 and 18 type [137]. Several other tumor-specific antigens, namely VEGF, carcinoembryonic antigen or CEA, EGFR, and HER-2/neu, are presently under exploration [138]. Furthermore, from oncogenic viruses, derived antigens, namely E6 and E7 oncoproteins associated with HPV, are documented as effectual targets for HNSCC patients [139].

6. Oral cancer

Oral malignancies are another debilitating clinical manifestation of human cancers globally and are accredited with survival rate of nearly 50%. The oral squamous cell carcinoma or OSCC is solely responsible for 90% of all the documented malignancies in the mouth, lips, and tongue [140]. Several risk factors associated with the oral malignancies include smoking, consumption of alcohol, viral infection, and chewing of betel nuts [141,142]. Technological advancements in clinical management of OSCC including surgical, chemo-, and/or radio-therapeutical interventions have proved inefficient in enhancing the posttreatment of life span of patients and are further reported to be instigator of severe side effects. The survival of patients in cases of oral malignancies like in other cases is dependent on known characteristics which include tumor grade and its invasiveness. However, recently, other biomarkers have also been reported with substantial expression in the clinical setting of OSCC [143,144]. Recently, immunotherapeutic interventions have proved their efficacy in OSCC and noteworthy is the neo-adjuvant-based interventions prior to surgery [145]. With the success of OSCC-based immunotherapy in the treatment of OSCC, it has become important to select parameters for selection of patients benefiting from the stated therapeutical strategy. To tackle the metastatic melanoma, the exploration of checkpoint inhibitors of immune system appears to be rationale for development of immune-based therapeutic interventions. Intriguingly, clinical trials in recent times have demonstrated the efficacy of PD-1 inhibitors for therapeutic management of metastatic patients of HNSCC. Nevertheless, there still exist certain issues that are still impeding the potential exploration of immunotherapeutics.

6.1 Immune checkpoint inhibitors

Inhibitors of immune checkpoints are now being pivotal for their role in management of oral potentially malignant disorders (OPMDs) where the investigation is based on identification of various biomarkers involved in prediction of OSCC-related transformation. It was elucidated PD-1 in association with its ligands are explicitly present on the surface of lymphocytes infiltrating the tissues in oral lichen planus (OLP) which plausibly be linked with tolerance toward inflammation of oral mucosa [146,147]. Contrastingly, the expression of PD-1 was relatively lesser in CD8+ lymphocytes during preliminary stages of tongue-associated OSCC [148]. Several researchers have further outlined the expression of PD-L1 on epithelial and/or subepithelial dysplastic lesions which can intrinsically evade the host's immune response and mediate the impediment of PD-1/PD-L1 pathway which in turn halts the transformation of malignant cells during OPMDs [149,150].

The OSCC tumor microenvironment or TME favors the characteristic adaptive response instigated by particular antigens [151]. In a study conducted by Ngamphaiboon and group, it was shown that nearly 84% of OSCC patients of a cohort exhibited positivity toward PD-L1 expression [151]. PD-1/PD-L1 in OSCC patients is also reported to correlate with elevated number of lymphocytes infiltrating the tumors [152–154]. Intriguingly, a correlation was demonstrated in a report suggesting augmented expression of both $CD4^+$ and $CD8^+$ lymphocytes was seen in tongue SCC [155]. The poor prognosis of OSCC is associated with expression of CD4 subset ofT lymphocytes comparatively than CD25; however, combined expression of PD-1/PD-L1 has been reported to reduce 5-year disease-related survival rate [156]. CD4 expression was linked with poor prognosis in OSCC compared to CD25, which has a more favorable outcome, and combined expression of substantially declined the 5-year disease-specific survival rate [156]. Furthermore, increase in PD-L1 expression frequently is related with the poor clinical observations [157,158]. Regulatory T cells have also been shown to play imperative role during suppression of tumor growth by impeding the proliferation of T cells and concomitant secretion of immunosuppressive cytokines [159]. A profound expression of cytosolic PD-L1 was further reported in circulating tumor cells which obstructed the T-cell-mediated inflammatory response [160–162].

Several chemotherapeutical candidates are still undergoing extensive exploration with intent of targeted therapies with curtailed side effects. Efficacy of inhibitors of immune checkpoints and chemo-therapeutic modality with PD-L1-mediated investigations have been undertaken in patients with OSCC [163,164]. Pembrolizumab is an FDA-approved anti-PD-1 inhibitor for patients with metastatic OSCC comparable to the tolerability and anticancer efficacy cabazitaxel [51,92]. Nivolumab is further reported for its efficacy in enhancing the rate of survival comparatively with standard therapeutical options with reduced toxic effects [39,97]. The role of PD-1/PD-L1 is presently underway in combinatorial approach radio-therapeutics. PD-L1 relates to sensitivity toward radio-therapeutics whereas survivin and c-Met relate to resistance toward radio-therapeutics [165].

7. Conclusion

Treatment of cancer is still a challenge in the field of medical science; however, the treatment modalities varying from surgery to chemotherapy as well as radiotherapy are yielding ambivalent outcomes. In this regard, novel immunotherapeutic approaches could be viewed as a savior in the never-ending battle against cancers. This chapter concluded the reviewed literature on available and novel immunotherapeutic interventions against HPV-associated malignancies. Antibodies like pembrolizumab designed against immune checkpoint inhibitors have demonstrated promising anticancer efficacy in patients suffering from HPV16 and PD-L1-positive cancers. However, these inhibitors are used along with radiotherapy, chemotherapy, or therapeutic agents. Amidst various prophylactic or therapeutic agents targeting HPV16, viz. VGX-3100 and ADXS11-001, are in final clinical trials, and it is anticipated that they will certainly get clinically approved. Nevertheless, majority of the therapeutic vaccines are still in preliminary trial phases. However, to achieve the success of cell-based therapies, successful approaches to tackle the immune evasive efficiency in tumor microenvironment and ameliorated trafficking of genetically tailored T cells into tumor tissues should be considered as important determinants. Owing to the complexity within the nature of HPV-positive carcinomas, various combination immunotherapies are highly needful.

References

[1] D.M. Parkin, The global health burden of infection-associated cancers in the year 2002, Int. J. Cancer 118 (12) (2006) 3030–3044.
[2] H.U. Bernard, R.D. Burk, Z. Chen, K. van Doorslaer, H. zur Hausen, E.M. de Villiers, Classification of papillomaviruses (PVs) based on 189 PV types and proposal of taxonomic amendments, Virology 401 (1) (2010) 70–79.
[3] M. Ling, M. Kanayama, R. Roden, T.C. Wu, Preventive and therapeutic vaccines for human papillomavirus-associated cervical cancers, J. Biomed. Sci. 7 (5) (2000) 341–356.
[4] N. Muñoz, F.X. Bosch, S. de Sanjosé, R. Herrero, X. Castellsagué, K.V. Shah, P.J. Snijders, C.J. Meijer, International Agency for Research on Cancer Multicenter Cervical Cancer Study Group, Epidemiologic classification of human papillomavirus types associated with cervical cancer, N. Engl. J. Med. 348 (6) (2003) 518–527.
[5] M. Schiffman, G. Clifford, F.M. Buonaguro, Classification of weakly carcinogenic human papillomavirus types: addressing the limits of epidemiology at the borderline, Infect. Agents Cancer 4 (2009) 8.
[6] P. Guan, R. Howell-Jones, N. Li, L. Bruni, S. de Sanjosé, S. Franceschi, G.M. Clifford, Human papillomavirus types in 115,789 HPV-positive women: a meta-analysis from cervical infection to cancer, Int. J. Cancer 131 (10) (2012) 2349–2359.
[7] A.K. Chaturvedi, Beyond cervical cancer: burden of other HPV-related cancers among men and women, J. Adolescent Health 46 (4 Suppl) (2010) S20–S26.

[8] Cancer attributable to infections, International Agency for Research on Cancer (IARC). (2018). [Internet]. GLOBOCAN. Available online at: https://gco.iarc.fr/causes/infections/tools-pie?mode=2&sex=2&population=hdi&continent=0&country=0&cancer=0&key=attr_cases&lock_scale=0&pie_mode=1&nb_results=5 from https://gco.iarc.fr/ (Accessed 2020).

[9] S. de Sanjosé, M. Diaz, X. Castellsagué, G. Clifford, L. Bruni, N. Muñoz, F.X. Bosch, Worldwide prevalence and genotype distribution of cervical human papillomavirus DNA in women with normal cytology: a meta-analysis, Lancet Infect. Dis. 7 (7) (2007) 453–459.

[10] A. Siddiqi, F. Webb, C. Smotherman, S. Shuja, A. Mubeen, K. Lai, Prevalence of epithelial abnormalities and high-risk human papilloma virus in cervicovaginal Pap smears of population subgroups as a guide toward evidence-based best practice, Diagn. Cytopathol. 47 (7) (2019) 648–652.

[11] C. Kabalika, D. Mulenga, M.L. Mazaba, S. Siziya, Acceptance of cervical cancer screening and its correlates among women of a peri-urban high-density residential area in Ndola, Zambia, Int. J. MCH AIDS 7 (1) (2018) 17–27.

[12] L. Bruni, M. Diaz, X. Castellsagué, E. Ferrer, F.X. Bosch, S. de Sanjosé, Cervical human papillomavirus prevalence in 5 continents: meta-analysis of 1 million women with normal cytological findings, J Infect Dis 202 (12) (2010) 1789–1799.

[13] K. Vinodhini, S. Shanmughapriya, B.C. Das, K. Natarajaseenivasan, Prevalence and risk factors of HPV infection among women from various provinces of the world, Arch. Gynecol. Obstet. 285 (3) (2012) 771–777.

[14] S. Alizon, C.L. Murall, I.G. Bravo, Why human papillomavirus acute infections matter, Viruses 9 (10) (2017) 293.

[15] S.C. Tan, M.P. Ismail, D.R. Duski, N.H. Othman, R. Ankathil, Prevalence and type distribution of human papillomavirus (HPV) in Malaysian women with and without cervical cancer: an updated estimate, Biosci. Rep. 38 (2) (2018), BSR20171268.

[16] S.K. Kjaer, B. Chackerian, A.J. van den Brule, E.I. Svare, G. Paull, J.M. Walbomers, J.T. Schiller, J.E. Bock, M.E. Sherman, D.R. Lowy, C.L. Meijer, High-risk human papillomavirus is sexually transmitted: evidence from a follow-up study of virgins starting sexual activity (intercourse), Cancer Epidemiol. Biomarkers Prev. 10 (2) (2001) 101–106.

[17] D. Wolday, M. Derese, S. Gebressellassie, B. Tsegaye, W. Ergete, Y. Gebrehiwot, O. Caplan, D.G. Wolf, S. Maayan, HPV genotype distribution among women with normal and abnormal cervical cytology presenting in a tertiary gynecology referral Clinic in Ethiopia, Infect. Agents Cancer 13 (2018) 28.

[18] C. Johnson, N. Obanor, A. DeWeese, Human papillomavirus and cancer in men, Health Sci. J. 10 (6) (2016) 1.

[19] A.N. Burchell, R.L. Winer, S. de Sanjosé, E.L. Franco, Chapter 6: Epidemiology and transmission dynamics of genital HPV infection, Vaccine 24 (Suppl 3) (2006) S3/52–S3/61.

[20] E.F. Dunne, C.M. Nielson, K.M. Stone, L.E. Markowitz, A.R. Giuliano, Prevalence of HPV infection among men: A systematic review of the literature, J Infect Dis 194 (8) (2006) 1044–1057.

[21] A. Nyitray, C.M. Nielson, R.B. Harris, R. Flores, M. Abrahamsen, E.F. Dunne, A.R. Giuliano, Prevalence of and risk factors for anal human papillomavirus infection in heterosexual men, J Infect Dis 197 (12) (2008) 1676–1684.

[22] E.D. Moreira Jr., A.R. Giuliano, J. Palefsky, C.A. Flores, S. Goldstone, D. Ferris, R.J. Hillman, H. Moi, M.H. Stoler, B. Marshall, S. Vuocolo, D. Guris, R.M. Haupt, Incidence, clearance, and disease progression of genital human papillomavirus infection in heterosexual men, J Infect Dis 210 (2) (2014) 192–199.

[23] HPV Study group in men from Brazil, USA and Mexico, Human papillomavirus infection in men residing in Brazil, Mexico, and the USA, Salud Publica Mex. 50 (5) (2008) 408–418.
[24] N. Moore, S. Lyle, Quiescent, slow-cycling stem cell populations in cancer: a review of the evidence and discussion of significance, J. Oncol. 2011 (2011) 396076.
[25] P.L. Stern, S.H. van der Burg, I.N. Hampson, T.R. Broker, A. Fiander, C.J. Lacey, H.C. Kitchener, M.H. Einstein, Therapy of human papillomavirus-related disease, Vaccine 30 (Suppl. 5) (2012) F71–F82.
[26] D. Milinovic, D. Kalafatic, D. Babic, L.B. Oreskovic, H.L. Grsic, S. Oreskovic, Minimally invasive therapy of cervical intraepithelial neoplasia for fertility preservation, Pathol. Oncol. Res.: POR 15 (3) (2009) 521–525.
[27] Cancer Genome Atlas Research Network, Analysis Working Group: Asan University, BC Cancer Agency, Brigham and Women's Hospital, Broad Institute, Brown University, Case Western Reserve University, Dana-Farber Cancer Institute, Duke University, Greater Poland Cancer Centre, Harvard Medical School, Institute for Systems Biology, KU Leuven, Mayo Clinic, Memorial Sloan Kettering Cancer Center, National Cancer Institute, Nationwide Children's Hospital, Stanford University, University of Alabama, University of Michigan, Project Team: National Institutes of Health, Integrated genomic characterization of oesophageal carcinoma, Nature 541 (7636) (2017) 169–175.
[28] P.G. Rose, Combined-modality therapy of locally advanced cervical cancer, J. Clin. Oncol. Off. J. Am. Soc. Clin. Oncol. 21 (10 Suppl) (2003) 211s–217s.
[29] D.H. Moore, J.A. Blessing, R.P. McQuellon, H.T. Thaler, D. Cella, J. Benda, D.S. Miller, G. Olt, S. King, J.F. Boggess, T.F. Rocereto, Phase III study of cisplatin with or without paclitaxel in stage IVB, recurrent, or persistent squamous cell carcinoma of the cervix: a gynecologic oncology group study, J. Clin. Oncol. Off. J. Am. Soc. Clin. Oncol. 22 (15) (2004) 3113–3119.
[30] B.J. Monk, M.W. Sill, D.S. McMeekin, D.E. Cohn, L.M. Ramondetta, C.H. Boardman, J. Benda, D. Cella, Phase III trial of four cisplatin-containing doublet combinations in stage IVB, recurrent, or persistent cervical carcinoma: a Gynecologic Oncology Group study, J. Clin. Oncol. Off. J. Am. Soc. Clin. Oncol. 27 (28) (2009) 4649–4655.
[31] K.S. Tewari, M.W. Sill, H.J. Long 3rd, R.T. Penson, H. Huang, L.M. Ramondetta, L.M. Landrum, A. Oaknin, T.J. Reid, M.M. Leitao, H.E. Michael, B.J. Monk, Improved survival with bevacizumab in advanced cervical cancer, N. Engl. J. Med. 370 (8) (2014) 734–743.
[32] B.S. Gill, M.E. Bernard, J.F. Lin, G.K. Balasubramani, M.S. Rajagopalan, P. Sukumvanich, T.C. Krivak, A.B. Olawaiye, J.L. Kelley, S. Beriwal, Impact of adjuvant chemotherapy with radiation for node-positive vulvar cancer: A national cancer data base (NCDB) analysis, Gynecol. Oncol. 137 (3) (2015) 365–372.
[33] National Comprehensive Cancer Network, NCCN Clinical Practice Guidelines in Oncology (NCCN Guidelines®). Vulvar cancer (squamous cell carcinoma). Version 1.2017, 2016. https://www.nccn.org/professionals/physician_gls/f_guidelines.asp. (Accessed 30 November 2016).
[34] P.J. Eifel, A.H. Klopp, J.S. Berek, P.A. Konstantinopoulos, Cancer of the cervix, vagina, and vulva, in: DeVita, Hellman, and Rosenberg's Cancer: Principles & Practice of Oncology, Wolters Kluwer Health Pharma Solutions (Europe) Ltd, 2018, pp. 1172–1210.
[35] R. Samant, B. Lau, E. Choan, T. Le, T. Tam, Primary vaginal cancer treated with concurrent chemoradiation using Cis-platinum, Int. J. Radiat. Oncol. Biol. Phys. 69 (3) (2007) 746–750.
[36] A.A. Forastiere, H. Goepfert, M. Maor, T.F. Pajak, R. Weber, W. Morrison, B. Glisson, A. Trotti, J.A. Ridge, C. Chao, G. Peters, D.J. Lee, A. Leaf, J. Ensley, J. Cooper, Concurrent

chemotherapy and radiotherapy for organ preservation in advanced laryngeal cancer, N. Engl. J. Med. 349 (22) (2003) 2091–2098.
[37] C. Taneja, H. Allen, R.J. Koness, K. Radie-Keane, H.J. Wanebo, Changing patterns of failure of head and neck cancer, Arch. Otolaryngol. Head Neck Surg. 128 (3) (2002) 324–327.
[38] S. Baxi, M. Fury, I. Ganly, S. Rao, D.G. Pfister, Ten years of progress in head and neck cancers, Journal of the National Comprehensive Cancer Network: JNCCN 10 (7) (2012) 806–810.
[39] R.L. Ferris, G. Blumenschein Jr., J. Fayette, J. Guigay, A.D. Colevas, L. Licitra, K. Harrington, S. Kasper, E.E. Vokes, C. Even, F. Worden, N.F. Saba, L.C. Iglesias Docampo, R. Haddad, T. Rordorf, N. Kiyota, M. Tahara, M. Monga, M. Lynch, W.J. Geese, M.L. Gillison, Nivolumab for recurrent squamous-cell carcinoma of the head and neck, N. Engl. J. Med. 375 (19) (2016) 1856–1867.
[40] T.Y. Seiwert, B. Burtness, R. Mehra, J. Weiss, R. Berger, J.P. Eder, K. Heath, T. McClanahan, J. Lunceford, C. Gause, J.D. Cheng, L.Q. Chow, Safety and clinical activity of pembrolizumab for treatment of recurrent or metastatic squamous cell carcinoma of the head and neck (KEYNOTE-012): an open-label, multicentre, phase 1b trial, Lancet Oncol. 17 (7) (2016) 956–965.
[41] US Food and Drug, FDA approves pembrolizumab for first-line treatment of head and neck squamous cell carcinoma, 2019, Available from: https://www.fda.gov/drugs/resources-information-approved-drugs/fda-approves-pembrolizumab-first-line-treatment-head-and-necksquamous- cell-carcinoma [Last accessed on 12 Nov 2019].
[42] J. Hamanishi, M. Mandai, I. Konishi, Immune checkpoint inhibition in ovarian cancer, Int. Immunol. 28 (7) (2016) 339–348.
[43] D. Pardoll, Cancer and the immune system: basic concepts and targets for intervention, Semin. Oncol. 42 (4) (2015) 523–538.
[44] IAfRo, Cancer Fact Sheets: Cervical Cancer, 2012, Available online: http://gco.iarc.fr/today/data/pdf/factsheets/cancers/cancer-fact-sheets-16.pdf (Accessed on 20 July 2019).
[45] W.J. Koh, N.R. Abu-Rustum, S. Bean, K. Bradley, S.M. Campos, K.R. Cho, H.S. Chon, C. Chu, R. Clark, D. Cohn, et al., NCCN Guidelines Panel Disclosures EmilyWyse Patient Advocate NCCN Guidelines Version 2.2018 Cervical Cancer, 2018, Available online: https://www.nccn.org/professionals/physician_gls/pdf/cervical.pdf (Accessed on 20 July 2019).
[46] X. Ji, K. Song, T. Lv, Y. Diao, N. Xia, Q. Yao, Nomogram for predicting the overall survival of cervical cancer patients with nonsurgical treatment, 2020. https://doi.org/10.21203/rs.3.rs-52519/v1.
[47] R.W. Naumann, A. Hollebecque, T. Meyer, M.J. Devlin, A. Oaknin, J. Kerger, J.M. López-Picazo, J.P. Machiels, J.P. Delord, T. Evans, V. Boni, E. Calvo, S.L. Topalian, T. Chen, I. Soumaoro, B. Li, J. Gu, R. Zwirtes, K.N. Moore, Safety and efficacy of nivolumab monotherapy in recurrent or metastatic cervical, vaginal, or vulvar carcinoma: results from the phase I/II CheckMate 358 trial, J. Clin. Oncol. Off. J. Am. Soc. Clin. Oncol. 37 (31) (2019) 2825–2834.
[48] A.D. Santin, W. Deng, M. Frumovitz, N. Buza, S. Bellone, W. Huh, S. Khleif, H.A. Lankes, E.S. Ratner, R.E. O'Cearbhaill, A.A. Jazaeri, M. Birrer, Phase II evaluation of nivolumab in the treatment of persistent or recurrent cervical cancer (NCT02257528/NRG-GY002), Gynecol. Oncol. 157 (1) (2020) 161–166.
[49] A. Hollebecque, T. Meyer, K.N. Moore, J.P.H. Machiels, J. De Greve, J.M. López-Picazo, S.L. Topalian, An open-label, multicohort, phase I/II study of nivolumab in patients with virus-associated tumors (CheckMate 358): Efficacy and safety in recurrent or metastatic (R/M) cervical, vaginal, and vulvar cancers, J. Clin. Oncol. Off. J. Am. Soc. Clin. Oncol. 35 (15 Suppl) (2017) 2825–2834.

[50] H.C. Chung, W. Ros, J.P. Delord, R. Perets, A. Italiano, R. Shapira-Frommer, L. Manzuk, S.A. Piha-Paul, L. Xu, S. Zeigenfuss, S.K. Pruitt, A. Leary, Efficacy and safety of pembrolizumab in previously treated advanced cervical cancer: results from the phase II KEYNOTE-158 study, J. Clin. Oncol. Off. J. Am. Soc. Clin. Oncol. 37 (17) (2019) 1470–1478.
[51] B. Burtness, K.J. Harrington, R. Greil, D. Soulières, M. Tahara, G. de Castro Jr., A. Psyrri, N. Basté, P. Neupane, Å. Bratland, T. Fuereder, B. Hughes, R. Mesía, N. Ngamphaiboon, T. Rordorf, W.Z. Wan Ishak, R.L. Hong, R. González Mendoza, A. Roy, Y. Zhang, KEYNOTE-048 Investigators, Pembrolizumab alone or with chemotherapy versus cetuximab with chemotherapy for recurrent or metastatic squamous cell carcinoma of the head and neck (KEYNOTE-048): a randomised, open-label, phase 3 study, Lancet (London, England) 394 (10212) (2019) 1915–1928.
[52] J.F. Grau, L. Farinas-Madrid, A. Oaknin, A randomized phase III trial of platinum chemotherapy plus paclitaxel with bevacizumab and atezolizumab versus platinum chemotherapy plus paclitaxel and bevacizumab in metastatic (stage IVB), persistent, or recurrent carcinoma of the cervix: the BEATcc study (ENGOT-Cx10/GEICO 68-C/JGOG1084/GOG-3030), Int. J. Gynecol. Cancer 30 (1) (2020) 139–143.
[53] Y. Zhao, W. Yang, Y. Huang, R. Cui, X. Li, B. Li, Evolving roles for targeting CTLA-4 in cancer immunotherapy, Cell. Physiol. Biochem. 47 (2) (2018) 721–734.
[54] D. Cunningham, N.C. Tebbutt, I. Davidenko, A.M. Murad, S.E. Al-Batran, D.H. Ilson, D.V.T. Catenacci, Phase III, randomized, double-blind, multicenter, placebo (P)-controlled trial of rilotumumab (R) plus epirubicin, cisplatin and capecitabine (ECX) as first-line therapy in patients (pts) with advanced MET-positive (pos) gastric or gastroesophageal junction (G/GEJ) cancer: RILOMET-1 study, J. Clin. Oncol. Off. J. Am. Soc. Clin. Oncol. 33 (15 Suppl) (2015) 4000.
[55] J. Mayadev, W.E. Brady, Y.G. Lin, D.M. Da Silva, H.A. Lankes, P.M. Fracasso, R.J. Schilder, A phase I study of sequential ipilimumab in the definitive treatment of node positive cervical cancer: GOG 9929, J. Clin. Oncol. Off. J. Am. Soc. Clin. Oncol. 35 (15 Suppl) (2017) 5526.
[56] M.K. Callahan, K. Odunsi, M. Sznol, J.J. Nemunaitis, P.A. Ott, P.M. Dillon, J.D. Wolchok, Phase 1 study to evaluate the safety and tolerability of MEDI4736 (durvalumab, DUR)+ tremelimumab (TRE) in patients with advanced solid tumors, J. Clin. Oncol. Off. J. Am. Soc. Clin. Oncol. 35 (15 Suppl) (2017) 3069.
[57] A. Chabeda, R. Yanez, R. Lamprecht, A.E. Meyers, E.P. Rybicki, I.I. Hitzeroth, Therapeutic vaccines for high-risk HPV-associated diseases, Papillomavirus Res. (Amsterdam, Netherlands) 5 (2018) 46–58.
[58] K. Lin, K. Doolan, C.F. Hung, T.C. Wu, Perspectives for preventive and therapeutic HPV vaccines, J. Formosan Med. Assoc. Taiwan 109 (1) (2010) 4–24.
[59] W.K. Huh, D.S. Dizon, M.A. Powell, C.A. Leath, L.M. Landrum, E. Tanner, C. Aghajanian, ADXS11-001 immunotherapy in squamous or non-squamous persistent/recurrent metastatic cervical cancer: Results from stage I of the phase II GOG/NRG0265 study, J. Clin. Oncol. Off. J. Am. Soc. Clin. Oncol. 34 (15 Suppl) (2016) 5516.
[60] P. Basu, A. Mehta, M. Jain, S. Gupta, R.V. Nagarkar, S. John, R. Petit, A randomized phase 2 study of ADXS11-001 Listeria monocytogenes-Listeriolysin O Immunotherapy with or without cisplatin in treatment of advanced cervical cancer, Int. J. Gynecol. Cancer 28 (4) (2018) 764–772.
[61] H. Safran, K.L. Leonard, T.A. DiPetrillo, A. Klipfel, S. Schechter, N. Oldenburg, L. Rajdev, ADXS11-001 Lm-LLO immunotherapy, mitomycin, 5-fluorouracil (5-FU) and intensity-modulated radiation therapy (IMRT) for anal cancer, J. Clin. Oncol. Off. J. Am. Soc. Clin. Oncol. 35 (15 Suppl) (2017), e15072.
[62] E.E. Cohen, K.N. Moore, B.M. Slomovitz, C.H. Chung, M.L. Anderson, S.R. Morris, B. Burtness, Phase I/II study of ADXS11-001 or MEDI4736 immunotherapies alone

and in combination, in patients with recurrent/metastatic cervical or human papillomavirus (HPV)-positive head and neck cancer, J. Immunother. Cancer 3 (suppl 2) (2015). P147.

[63] E. Massarelli, W. William, F. Johnson, M. Kies, R. Ferrarotto, M. Guo, L. Feng, J.J. Lee, H. Tran, Y.U. Kim, C. Haymaker, C. Bernatchez, M. Curran, T. Zecchini Barrese, J. Rodriguez Canales, I. Wistuba, L. Li, J. Wang, S.H. van der Burg, C.J. Melief, B. Glisson, Combining immune checkpoint blockade and tumor-specific vaccine for patients with incurable human papillomavirus 16-related cancer: a phase 2 clinical trial, JAMA Oncol. 5 (1) (2019) 67–73.

[64] W.R. Gerritsen, C.J. Melief, M. Welters, I. Vergote, J.R. Kroep, G. Kenter, S. van der Burg, Association of T cell responses after vaccination with HPV16 long peptides for late stage cervical cancer with prolonged survival, J. Clin. Oncol. Off. J. Am. Soc. Clin. Oncol. 35 (15 Suppl) (2017) 5525.

[65] P. Vici, L. Pizzuti, L. Mariani, G. Zampa, D. Santini, L. Di Lauro, T. Gamucci, C. Natoli, P. Marchetti, M. Barba, M. Maugeri-Saccà, D. Sergi, F. Tomao, E. Vizza, S. Di Filippo, F. Paolini, G. Curzio, G. Corrado, A. Michelotti, G. Sanguineti, A. Venuti, Targeting immune response with therapeutic vaccines in premalignant lesions and cervical cancer: hope or reality from clinical studies, Expert Rev. Vaccines 15 (10) (2016) 1327–1336.

[66] A. Yang, E. Farmer, T.C. Wu, C.F. Hung, Perspectives for therapeutic HPV vaccine development, J. Biomed. Sci. 23 (1) (2016) 75.

[67] Y.J. Choi, S.Y. Hur, T.J. Kim, S.R. Hong, J.K. Lee, C.H. Cho, K.S. Park, J.W. Woo, Y.C. Sung, Y.S. Suh, J.S. Park, A phase II, prospective, randomized, multicenter, open-label study of GX-188E, an HPV DNA vaccine, in patients with cervical intraepithelial neoplasia 3, Clin. Cancer Res. 26 (7) (2020) 1616–1623.

[68] L. Yu, Z. Hu, C. Gao, B. Feng, L. Wang, X. Tian, H. Wang, Deletion of HPV18 E6 and E7 genes using dual sgRNA-directed CRISPR/Cas9 inhibits growth of cervical cancer cells, Int. J. Clin. Exp. Med. 10 (6) (2017) 9206–9213.

[69] Z. Hu, W. Ding, D. Zhu, L. Yu, X. Jiang, X. Wang, C. Zhang, L. Wang, T. Ji, D. Liu, D. He, X. Xia, T. Zhu, J. Wei, P. Wu, C. Wang, L. Xi, Q. Gao, G. Chen, R. Liu, H. Wang, TALEN-mediated targeting of HPV oncogenes ameliorates HPV-related cervical malignancy, J. Clin. Invest. 125 (1) (2015) 425–436.

[70] W. Ding, Z. Hu, D. Zhu, X. Jiang, L. Yu, X. Wang, C. Zhang, L. Wang, T. Ji, K. Li, D. He, X. Xia, D. Liu, J. Zhou, D. Ma, H. Wang, Zinc finger nucleases targeting the human papillomavirus E7 oncogene induce E7 disruption and a transformed phenotype in HPV16/18-positive cervical cancer cells, Clin. Cancer Res. 20 (24) (2014) 6495–6503.

[71] S. Stevanović, L.M. Draper, M.M. Langhan, T.E. Campbell, M.L. Kwong, J.R. Wunderlich, M.E. Dudley, J.C. Yang, R.M. Sherry, U.S. Kammula, N.P. Restifo, S.A. Rosenberg, C.S. Hinrichs, Complete regression of metastatic cervical cancer after treatment with human papillomavirus-targeted tumor-infiltrating T cells, J. Clin. Oncol. Off. J. Am. Soc. Clin. Oncol. 33 (14) (2015) 1543–1550.

[72] C.A. Ramos, N. Narala, G.M. Vyas, A.M. Leen, U. Gerdemann, E.M. Sturgis, M.L. Anderson, B. Savoldo, H.E. Heslop, M.K. Brenner, C.M. Rooney, Human papillomavirus type 16 E6/E7-specific cytotoxic T lymphocytes for adoptive immunotherapy of HPV-associated malignancies, J. Immunotherapy (Hagerstown, Md.: 1997) 36 (1) (2013) 66–76.

[73] S. Stevanović, S.R. Helman, J.R. Wunderlich, M.M. Langhan, S.L. Doran, M. Kwong, R. Somerville, C.A. Klebanoff, U.S. Kammula, R.M. Sherry, J.C. Yang, S.A. Rosenberg, C.S. Hinrichs, A Phase II study of tumor-infiltrating lymphocyte therapy for human papillomavirus-associated epithelial cancers, Clin. Cancer Res. 25 (5) (2019) 1486–1493.

[74] E.N. Baruch, A.L. Berg, M.J. Besser, J. Schachter, G. Markel, Adoptive T cell therapy: An overview of obstacles and opportunities, Cancer 123 (S11) (2017) 2154–2162. Poorebrahim M, Sadeghi S, Fakhr E Abazari MF, Poortahmasebi V, Kheirollahi A, Askari H, Rajabzadeh A, Rastegarpanah M, Linē A, & Cid-Arregui A. Production of CAR T-cells by GMP-grade lentiviral vectors: latest advances and future prospects. Crit Rev Clin Lab Sci, 56(6), 393–419.
[75] Y. Ping, C. Liu, Y. Zhang, T-cell receptor-engineered T cells for cancer treatment: current status and future directions, Protein Cell 9 (3) (2018) 254–266.
[76] M.A. de Witte, M. Coccoris, M.C. Wolkers, M.D. van den Boom, E.M. Mesman, J.Y. Song, M. van der Valk, J.B. Haanen, T.N. Schumacher, Targeting self-antigens through allogeneic TCR gene transfer, Blood 108 (3) (2006) 870–877.
[77] T.K. Hoffmann, C. Arsov, K. Schirlau, M. Bas, U. Friebe-Hoffmann, J.P. Klussmann, K. Scheckenbach, V. Balz, H. Bier, T.L. Whiteside, T cells specific for HPV16 E7 epitopes in patients with squamous cell carcinoma of the oropharynx, Int. J. Cancer 118 (8) (2006) 1984–1991.
[78] M. Sharpe, N. Mount, Genetically modified T cells in cancer therapy: opportunities and challenges, Dis. Model. Mech. 8 (4) (2015) 337–350.
[79] E. Rosati, C.M. Dowds, E. Liaskou, E. Henriksen, T.H. Karlsen, A. Franke, Overview of methodologies for T-cell receptor repertoire analysis, BMC Biotechnol. 17 (1) (2017) 61.
[80] Y. Zhao, Z. Zheng, C.J. Cohen, L. Gattinoni, D.C. Palmer, N.P. Restifo, S.A. Rosenberg, R.A. Morgan, High-efficiency transfection of primary human and mouse T lymphocytes using RNA electroporation, Mol. Therapy: J. Amer. Soc. Gene Therapy 13 (1) (2006) 151–159.
[81] J.A. Williams, Vector design for improved DNA vaccine efficacy, safety and production, Vaccine 1 (3) (2013) 225–249.
[82] V. Gaspar, D. de Melo-Diogo, E. Costa, A. Moreira, J. Queiroz, C. Pichon, I. Correia, F. Sousa, Minicircle DNA vectors for gene therapy: advances and applications, Expert Opin. Biol. Ther. 15 (3) (2015) 353–379.
[83] J.E. Vargas, L. Chicaybam, R.T. Stein, A. Tanuri, A. Delgado-Cañedo, M.H. Bonamino, Retroviral vectors and transposons for stable gene therapy: advances, current challenges and perspectives, J. Transl. Med. 14 (1) (2016) 288.
[84] M.C. Milone, U. O'Doherty, Clinical use of lentiviral vectors, Leukemia 32 (7) (2018) 1529–1541.
[85] F. Bray, J. Ferlay, I. Soerjomataram, R.L. Siegel, L.A. Torre, A. Jemal, Global cancer statistics 2018: GLOBOCAN estimates of incidence and mortality worldwide for 36 cancers in 185 countries, CA Cancer J. Clin. 68 (6) (2018) 394–424.
[86] R.I. Haddad, D.M. Shin, Recent advances in head and neck cancer, N. Engl. J. Med. 359 (11) (2008) 1143–1154.
[87] T.B. Steinbichler, M. Lichtenecker, M. Anegg, D. Dejaco, B. Kofler, V.H. Schartinger, M.T. Kasseroler, B. Forthuber, A. Posch, H. Riechelmann, Persistent head and neck cancer following first-line treatment, Cancer 10 (11) (2018) 421.
[88] J.B. Vermorken, R. Mesia, F. Rivera, E. Remenar, A. Kawecki, S. Rottey, J. Erfan, D. Zabolotnyy, H.R. Kienzer, D. Cupissol, F. Peyrade, M. Benasso, I. Vynnychenko, D. De Raucourt, C. Bokemeyer, A. Schueler, N. Amellal, R. Hitt, Platinum-based chemotherapy plus cetuximab in head and neck cancer, N. Engl. J. Med. 359 (11) (2008) 1116–1127.
[89] R. Mandal, Y. Şenbabaoğlu, A. Desrichard, J.J. Havel, M.G. Dalin, N. Riaz, K.W. Lee, I. Ganly, A.A. Hakimi, T.A. Chan, L.G. Morris, The head and neck cancer immune landscape and its immunotherapeutic implications, JCI Insight 1 (17) (2016), e89829.
[90] R.L. Ferris, Immunology and immunotherapy of head and neck cancer, J. Clin. Oncol. Off. J. Am. Soc. Clin. Oncol. 33 (29) (2015) 3293–3304.

[91] L. Chow, R. Haddad, S. Gupta, A. Mahipal, R. Mehra, M. Tahara, R. Berger, J.P. Eder, B. Burtness, S.H. Lee, B. Keam, H. Kang, K. Muro, J. Weiss, R. Geva, C.C. Lin, H.C. Chung, A. Meister, M. Dolled-Filhart, K. Pathiraja, T.Y. Seiwert, Antitumor activity of pembrolizumab in biomarker-unselected patients with recurrent and/or metastatic head and neck squamous cell carcinoma: results from the phase Ib KEYNOTE-012 expansion cohort, J. Clin. Oncol. Off. J. Am. Soc. Clin. Oncol. 34 (32) (2016) 3838–3845.

[92] J. Bauml, T.Y. Seiwert, D.G. Pfister, F. Worden, S.V. Liu, J. Gilbert, N.F. Saba, J. Weiss, L. Wirth, A. Sukari, H. Kang, M.K. Gibson, E. Massarelli, S. Powell, A. Meister, X. Shu, J.D. Cheng, R. Haddad, Pembrolizumab for platinum- and cetuximab-refractory head and neck cancer: results from a single-arm, phase II study, J. Clin. Oncol. Off. J. Am. Soc. Clin. Oncol. 35 (14) (2017) 1542–1549.

[93] J.E. Bauman, E. Cohen, R.L. Ferris, D.J. Adelstein, D.M. Brizel, J.A. Ridge, B. O'Sullivan, B.A. Burtness, L.H. Butterfield, W.E. Carson, M.L. Disis, B.A. Fox, T.F. Gajewski, M.L. Gillison, J.W. Hodge, Q.T. Le, D. Raben, S.E. Strome, J. Lynn, S. Malik, Immunotherapy of head and neck cancer: Emerging clinical trials from a National Cancer Institute Head and Neck Cancer Steering Committee Planning Meeting, Cancer 123 (7) (2017) 1259–1271.

[94] B. Samra, E. Tam, B. Baseri, I. Shapira, Checkpoint inhibitors in head and neck cancer: current knowledge and perspectives, J. Invest. Med. 66 (7) (2018) 1023–1030.

[95] J.R. Brahmer, C.G. Drake, I. Wollner, J.D. Powderly, J. Picus, W.H. Sharfman, E. Stankevich, A. Pons, T.M. Salay, T.L. McMiller, M.M. Gilson, C. Wang, M. Selby, J.M. Taube, R. Anders, L. Chen, A.J. Korman, D.M. Pardoll, I. Lowy, S.L. Topalian, Phase I study of single-agent anti-programmed death-1 (MDX-1106) in refractory solid tumors: safety, clinical activity, pharmacodynamics, and immunologic correlates, J. Clin. Oncol. Off. J. Am. Soc. Clin. Oncol. 28 (19) (2010) 3167–3175.

[96] S.L. Topalian, F.S. Hodi, J.R. Brahmer, S.N. Gettinger, D.C. Smith, D.F. McDermott, J.D. Powderly, R.D. Carvajal, J.A. Sosman, M.B. Atkins, P.D. Leming, D.R. Spigel, S.J. Antonia, L. Horn, C.G. Drake, D.M. Pardoll, L. Chen, W.H. Sharfman, R.A. Anders, J.M. Taube, M. Sznol, Safety, activity, and immune correlates of anti-PD-1 antibody in cancer, N. Engl. J. Med. 366 (26) (2012) 2443–2454.

[97] K.J. Harrington, R.L. Ferris, G. Blumenschein Jr., A.D. Colevas, J. Fayette, L. Licitra, S. Kasper, C. Even, E.E. Vokes, F. Worden, N.F. Saba, N. Kiyota, R. Haddad, M. Tahara, V. Grünwald, J.W. Shaw, M. Monga, M. Lynch, F. Taylor, M. DeRosa, J. Guigay, Nivolumab versus standard, single-agent therapy of investigator's choice in recurrent or metastatic squamous cell carcinoma of the head and neck (CheckMate 141): health-related quality-of-life results from a randomised, phase 3 trial, Lancet Oncol. 18 (8) (2017) 1104–1115.

[98] C.J. Melief, T. van Hall, R. Arens, F. Ossendorp, S.H. van der Burg, Therapeutic cancer vaccines, J. Clin. Invest. 125 (9) (2015) 3401–3412.

[99] J. Rice, C.H. Ottensmeier, F.K. Stevenson, DNA vaccines: precision tools for activating effective immunity against cancer. Nature reviews, Cancer 8 (2) (2008) 108–120.

[100] R. Rosales, M. López-Contreras, C. Rosales, J.R. Magallanes-Molina, R. Gonzalez-Vergara, J.M. Arroyo-Cazarez, A. Ricardez-Arenas, A. Del Follo-Valencia, S. Padilla-Arriaga, M.V. Guerrero, M.A. Pirez, C. Arellano-Fiore, F. Villarreal, Regression of human papillomavirus intraepithelial lesions is induced by MVA E2 therapeutic vaccine, Hum. Gene Ther. 25 (12) (2014) 1035–1049.

[101] E. Quoix, H. Lena, G. Losonczy, F. Forget, C. Chouaid, Z. Papai, R. Gervais, C. Ottensmeier, A. Szczesna, A. Kazarnowicz, J.T. Beck, V. Westeel, E. Felip, D. Debieuvre, A. Madroszyk, J. Adam, G. Lacoste, A. Tavernaro, B. Bastien, C. Halluard, J.M. Limacher, TG4010 immunotherapy and first-line chemotherapy for advanced non-small-cell

lung cancer (TIME): results from the phase 2b part of a randomised, double-blind, placebo-controlled, phase 2b/3 trial, Lancet Oncol. 17 (2) (2016) 212–223.
[102] L. Low, A. Mander, K. McCann, D. Dearnaley, T. Tjelle, I. Mathiesen, F. Stevenson, C.H. Ottensmeier, DNA vaccination with electroporation induces increased antibody responses in patients with prostate cancer, Hum. Gene Ther. 20 (11) (2009) 1269–1278.
[103] A.N. Meleshko, N.A. Petrovskaya, N. Savelyeva, K.P. Vashkevich, S.N. Doronina, N.V. Sachivko, Phase I clinical trial of idiotypic DNA vaccine administered as a complex with polyethylenimine to patients with B-cell lymphoma, Hum. Vaccin. Immunother. 13 (6) (2017) 1–6.
[104] A. Bolhassani, S. Safaiyan, S. Rafati, Improvement of different vaccine delivery systems for cancer therapy, Mol. Cancer 10 (2011) 3.
[105] M.E. Boursnell, E. Rutherford, J.K. Hickling, E.A. Rollinson, A.J. Munro, N. Rolley, C.S. McLean, L.K. Borysiewicz, K. Vousden, S.C. Inglis, Construction and characterisation of a recombinant vaccinia virus expressing human papillomavirus proteins for immunotherapy of cervical cancer, Vaccine 14 (16) (1996) 1485–1494.
[106] D. Pim, A. Storey, M. Thomas, P. Massimi, L. Banks, Mutational analysis of HPV-18 E6 identifies domains required for p53 degradation in vitro, abolition of p53 transactivation in vivo and immortalisation of primary BMK cells, Oncogene 9 (7) (1994) 1869–1876.
[107] C.L. Trimble, M.P. Morrow, K.A. Kraynyak, X. Shen, M. Dallas, J. Yan, L. Edwards, R.L. Parker, L. Denny, M. Giffear, A.S. Brown, K. Marcozzi-Pierce, D. Shah, A.M. Slager, A.J. Sylvester, A. Khan, K.E. Broderick, R.J. Juba, T.A. Herring, J. Boyer, M.L. Bagarazzi, Safety, efficacy, and immunogenicity of VGX-3100, a therapeutic synthetic DNA vaccine targeting human papillomavirus 16 and 18 E6 and E7 proteins for cervical intraepithelial neoplasia 2/3: a randomised, double-blind, placebo-controlled phase 2b trial, Lancet (London, England) 386 (10008) (2015) 2078–2088.
[108] C. Aggarwal, R. Cohen, M.P. Morrow, J. Bauml, G. Weinstein, J. Boyer, D. Weiner, Immunotherapy with VGX-3100 (HPV16 and HPV18 plasmids)+ INO-9012 (DNA encoding IL-12) in human papillomavirus (HPV) associated head and neck squamous cell carcinoma (HNSCCa): interim safety and immunogenicity results, J. Immunother. Cancer 3 (2) (2015) 1–2.
[109] W.F. Cheng, C.F. Hung, C.Y. Chai, K.F. Hsu, L. He, M. Ling, T.C. Wu, Tumor-specific immunity and antiangiogenesis generated by a DNA vaccine encoding calreticulin linked to a tumor antigen, J. Clin. Invest. 108 (5) (2001) 669–678.
[110] S.R. Best, S. Peng, C.M. Juang, C.F. Hung, D. Hannaman, J.R. Saunders, T.C. Wu, S.I. Pai, Administration of HPV DNA vaccine via electroporation elicits the strongest CD8+ T cell immune responses compared to intramuscular injection and intradermal gene gun delivery, Vaccine 27 (40) (2009) 5450–5459.
[111] R.D. Alvarez, W.K. Huh, S. Bae, L.S. Lamb Jr., M.G. Conner, J. Boyer, et al., A pilot study of pNGVL4a-CRT/E7(detox) for the treatment of patients with HPV16+ cervical intraepithelial neoplasia 2/3 (CIN2/3), Gynecol. Oncol. 140 (2) (2016) 245–252.
[112] A. Allen, C. Wang, L.J. Caproni, G. Sugiyarto, E. Harden, L.R. Douglas, P.J. Duriez, K. Karbowniczek, J. Extance, P.J. Rothwell, I. Orefo, J.P. Tite, F.K. Stevenson, C.H. Ottensmeier, N. Savelyeva, Linear doggybone DNA vaccine induces similar immunological responses to conventional plasmid DNA independently of immune recognition by TLR9 in a pre-clinical model, Cancer Immunology, Immunother.: CII 67 (4) (2018) 627–638.
[113] L.M. Kranz, M. Diken, H. Haas, S. Kreiter, C. Loquai, K.C. Reuter, M. Meng, D. Fritz, F. Vascotto, H. Hefesha, C. Grunwitz, M. Vormehr, Y. Hüsemann, A. Selmi, A.N. Kuhn, J. Buck, E. Derhovanessian, R. Rae, S. Attig, J. Diekmann, U. Sahin, Systemic RNA delivery to dendritic cells exploits antiviral defence for cancer immunotherapy, Nature 534 (7607) (2016) 396–401.

[114] U. Sahin, E. Derhovanessian, M. Miller, B.P. Kloke, P. Simon, M. Löwer, V. Bukur, A.D. Tadmor, U. Luxemburger, B. Schrörs, T. Omokoko, M. Vormehr, C. Albrecht, A. Paruzynski, A.N. Kuhn, J. Buck, S. Heesch, K.H. Schreeb, F. Müller, I. Ortseifer, Ö. Türeci, Personalized RNA mutanome vaccines mobilize poly-specific therapeutic immunity against cancer, Nature 547 (7662) (2017) 222–226.

[115] D. Chatzikleanthous, D.T. O'Hagan, R. Adamo, Lipid-based nanoparticles for delivery of vaccine adjuvants and antigens: toward multicomponent vaccines, Mol. Pharm. 18 (8) (2021) 2867–2888.

[116] P.J. de Vos van Steenwijk, M.I. van Poelgeest, T.H. Ramwadhdoebe, M.J. Löwik, D.M. Berends-van der Meer, C.E. van der Minne, N.M. Loof, L.F. Stynenbosch, L.M. Fathers, A.R. Valentijn, J. Oostendorp, E.M. Osse, G.J. Fleuren, L. Nooij, M.J. Kagie, B.W. Hellebrekers, C.J. Melief, M.J. Welters, S.H. van der Burg, G.G. Kenter, The long-term immune response after HPV16 peptide vaccination in women with low-grade pre-malignant disorders of the uterine cervix: a placebo-controlled phase II study, Cancer Immunology, Immunother.: CII 63 (2) (2014) 147–160.

[117] M.I. van Poelgeest, M.J. Welters, E.M. van Esch, L.F. Stynenbosch, G. Kerpershoek, E.L.V.P. van Meerten, S.H. van der Burg, HPV16 synthetic long peptide (HPV16-SLP) vaccination therapy of patients with advanced or recurrent HPV16-induced gynecological carcinoma, a phase II trial, J. Transl. Med. 11 (1) (2013) 88.

[118] M.J. Welters, T.C. van der Sluis, H. van Meir, N.M. Loof, V.J. van Ham, S. van Duikeren, S.J. Santegoets, R. Arens, M.L. de Kam, A.F. Cohen, M.I. van Poelgeest, G.G. Kenter, J.R. Kroep, J. Burggraaf, C.J. Melief, S.H. van der Burg, Vaccination during myeloid cell depletion by cancer chemotherapy fosters robust T cell responses, Sci. Transl. Med. 8 (334) (2016), 334ra52.

[119] V. Bechtold, P. Beard, K. Raj, Human papillomavirus type 16 E2 protein has no effect on transcription from episomal viral DNA, J. Virol. 77 (3) (2003) 2021–2028.

[120] A.A. McBride, The papillomavirus E2 proteins, Virology 445 (1–2) (2013) 57–79.

[121] N.H. Segal, D.W. Parsons, K.S. Peggs, V. Velculescu, K.W. Kinzler, B. Vogelstein, J.P. Allison, Epitope landscape in breast and colorectal cancer, Cancer Res. 68 (3) (2008) 889–892.

[122] J.S. Weber, G.T. Gibney, B. Yu, P. Cheng, A.J. Martinez, J. Kroeger, M.J. Schell, Survival, biomarker, and toxicity analysis of nivolumab (NIVO) in patients that progressed on ipilimumab (IPI), J. Clin. Oncol. Off. J. Am. Soc. Clin. Oncol. 33 (15 Suppl) (2015) 9055.

[123] G.L. Eiben, M.P. Velders, H. Schreiber, M.C. Cassetti, J.K. Pullen, L.R. Smith, W.M. Kast, Establishment of an HLA-A*0201 human papillomavirus type 16 tumor model to determine the efficacy of vaccination strategies in HLA-A*0201 transgenic mice, Cancer Res. 62 (20) (2002) 5792–5799.

[124] Y. Yoshitake, D. Fukuma, A. Yuno, M. Hirayama, H. Nakayama, T. Tanaka, M. Nagata, Y. Takamune, K. Kawahara, Y. Nakagawa, R. Yoshida, A. Hirosue, H. Ogi, A. Hiraki, H. Jono, A. Hamada, K. Yoshida, Y. Nishimura, Y. Nakamura, M. Shinohara, Phase II clinical trial of multiple peptide vaccination for advanced head and neck cancer patients revealed induction of immune responses and improved OS, Clin. Cancer Res. 21 (2) (2015) 312–321.

[125] T.L. Whiteside, Immunobiology and immunotherapy of head and neck cancer, Curr. Oncol. Rep. 3 (1) (2001) 46–55.

[126] P. Comoli, P. Pedrazzoli, R. Maccario, S. Basso, O. Carminati, M. Labirio, R. Schiavo, S. Secondino, C. Frasson, C. Perotti, M. Moroni, F. Locatelli, S. Siena, Cell therapy of stage IV nasopharyngeal carcinoma with autologous Epstein-Barr virus-targeted cytotoxic T lymphocytes, J. Clin. Oncol. Off. J. Am. Soc. Clin. Oncol. 23 (35) (2005) 8942–8949.

[127] A.E. Chang, Q. Li, G. Jiang, T.N. Teknos, D.B. Chepeha, C.R. Bradford, Generation of vaccine-primed lymphocytes for the treatment of head and neck cancer, Head Neck 25 (3) (2003) 198–209.
[128] W.C. To, B.G. Wood, J.C. Krauss, M. Strome, R.M. Esclamado, P. Lavertu, D. Dasko, J.A. Kim, G.E. Plautz, B.E. Leff, V. Smith, K. Sandstrom-Wakeling, S. Shu, Systemic adoptive T-cell immunotherapy in recurrent and metastatic carcinoma of the head and neck: a phase 1 study, Arch. Otolaryngol.—Head & Neck Surgery 126 (10) (2000) 1225–1231.
[129] E. Artusio, B. Hathaway, J. Stanson, T.L. Whiteside, Transfection of human monocyte-derived dendritic cells with native tumor DNA induces antigen-specific T-cell responses in vitro, Cancer Biol. Ther. 5 (12) (2006) 1624–1631.
[130] J.S. Moyer, J. Li, S. Wei, S. Teitz-Tennenbaum, A.E. Chang, Intratumoral dendritic cells and chemoradiation for the treatment of murine squamous cell carcinoma, J. Immunotherapy (Hagerstown, Md.: 1997) 31 (9) (2008) 885–895.
[131] A.A. Wu, K.J. Niparko, S.I. Pai, Immunotherapy for head and neck cancer, J. Biomed. Sci. 15 (3) (2008) 275–289.
[132] M. Eura, K. Ogi, K. Chikamatsu, K.D. Lee, K. Nakano, K. Masuyama, K. Itoh, T. Ishikawa, Expression of the MAGE gene family in human head-and-neck squamous-cell carcinomas, Int. J. Cancer 64 (5) (1995) 304–308.
[133] M.A. Kienstra, H.B. Neel, S.E. Strome, P. Roche, Identification of NY-ESO-1, MAGE-1, and MAGE-3 in head and neck squamous cell carcinoma, Head Neck 25 (6) (2003) 457–463.
[134] Y.T. Chen, M.J. Scanlan, U. Sahin, O. Türeci, A.O. Gure, S. Tsang, B. Williamson, E. Stockert, M. Pfreundschuh, L.J. Old, A testicular antigen aberrantly expressed in human cancers detected by autologous antibody screening, Proc. Natl. Acad. Sci. U. S. A. 94 (5) (1997) 1914–1918.
[135] C.E. Clark, R.H. Vonderheide, Cancer-testis antigens in tumor biology and immunotherapy, Cancer Biol. Ther. 5 (9) (2006) 1226–1227.
[136] V. Balz, K. Scheckenbach, K. Götte, U. Bockmühl, I. Petersen, H. Bier, Is the p53 inactivation frequency in squamous cell carcinomas of the head and neck underestimated? Analysis of p53 exons 2-11 and human papillomavirus 16/18 E6 transcripts in 123 unselected tumor specimens, Cancer Res. 63 (6) (2003) 1188–1191.
[137] A. Albers, K. Abe, J. Hunt, J. Wang, A. Lopez-Albaitero, C. Schaefer, W. Gooding, T.L. Whiteside, S. Ferrone, A. DeLeo, R.L. Ferris, Antitumor activity of human papillomavirus type 16 E7-specific T cells against virally infected squamous cell carcinoma of the head and neck, Cancer Res. 65 (23) (2005) 11146–11155.
[138] K. Devaraj, M.L. Gillison, T.C. Wu, Development of HPV vaccines for HPV-associated head and neck squamous cell carcinoma, Crit Rev Oral Biol Med 14 (5) (2003) 345–362.
[139] M.L. Gillison, W.M. Koch, R.B. Capone, M. Spafford, W.H. Westra, L. Wu, M.L. Zahurak, R.W. Daniel, M. Viglione, D.E. Symer, K.V. Shah, D. Sidransky, Evidence for a causal association between human papillomavirus and a subset of head and neck cancers, J. Natl. Cancer Inst. 92 (9) (2000) 709–720.
[140] A.K. Markopoulos, Current aspects on oral squamous cell carcinoma, Open Dent. J. 6 (2012) 126–130.
[141] C.Y. Chen, S.H. Chiou, C.Y. Huang, C.I. Jan, S.C. Lin, W.Y. Hu, S.H. Chou, C.J. Liu, J.F. Lo, Tid1 functions as a tumour suppressor in head and neck squamous cell carcinoma, J. Pathol. 219 (3) (2009) 347–355.
[142] Y.Y. Hsu, C.H. Bai, C.C. Wang, W.L. Chen, W.T. Wu, C.H. Lai, Health disparities of employees in taiwan with major cancer diagnosis from 2004 to 2015: a nation- and population-based analysis, Int. J. Environ. Res. Public Health 16 (11) (2019) 1982.

[143] A. Almangush, I. Heikkinen, A.A. Mäkitie, R.D. Coletta, E. Läärä, I. Leivo, T. Salo, Prognostic biomarkers for oral tongue squamous cell carcinoma: a systematic review and meta-analysis, Br. J. Cancer 117 (6) (2017) 856–866.
[144] C. Rivera, A.K. Oliveira, R. Costa, T. De Rossi, A.F. Paes Leme, Prognostic biomarkers in oral squamous cell carcinoma: a systematic review, Oral Oncol. 72 (2017) 38–47.
[145] J.D. Schoenfeld, G.J. Hanna, V.Y. Jo, B. Rawal, Y.H. Chen, P.S. Catalano, A. Lako, Z. Ciantra, J.L. Weirather, S. Criscitiello, A. Luoma, N. Chau, J. Lorch, J.I. Kass, D. Annino, L. Goguen, A. Desai, B. Ross, H.J. Shah, H.A. Jacene, R.I. Haddad, Neoadjuvant nivolumab or nivolumab plus ipilimumab in untreated oral cavity squamous cell carcinoma: a phase 2 open-label randomized clinical trial, JAMA Oncol. 6 (10) (2020) 1563–1570.
[146] P. Youngnak-Piboonratanakit, F. Tsushima, N. Otsuki, H. Igarashi, U. Machida, H. Iwai, Y. Takahashi, K. Omura, H. Yokozeki, M. Azuma, The expression of B7-H1 on keratinocytes in chronic inflammatory mucocutaneous disease and its regulatory role, Immunol. Lett. 94 (3) (2004) 215–222.
[147] G.H. Du, X.P. Qin, Q. Li, Y.M. Zhou, X.M. Shen, G.Y. Tang, The high expression level of programmed death-1 ligand 2 in oral lichen planus and the possible costimulatory effect on human T cells, J. Oral Pathol. Med. 40 (7) (2011) 525–532.
[148] F. Katou, H. Ohtani, Y. Watanabe, T. Nakayama, O. Yoshie, K. Hashimoto, Differing phenotypes between intraepithelial and stromal lymphocytes in early-stage tongue cancer, Cancer Res. 67 (23) (2007) 11195–11201.
[149] T. Yagyuu, K. Hatakeyama, M. Imada, M. Kurihara, Y. Matsusue, K. Yamamoto, C. Obayashi, T. Kirita, Programmed death ligand 1 (PD-L1) expression and tumor microenvironment: Implications for patients with oral precancerous lesions, Oral Oncol. 68 (2017) 36–43.
[150] G. Zhou, J. Zhang, X.W. Ren, J.Y. Hu, G.F. Du, X.Y. Xu, Increased B7-H1 expression on peripheral blood T cells in oral lichen planus correlated with disease severity, J. Clin. Immunol. 32 (4) (2012) 794–801.
[151] H. Quan, L. Fang, H. Pan, Z. Deng, S. Gao, O. Liu, Y. Wang, Y. Hu, X. Fang, Z. Yao, F. Guo, R. Lu, K. Xia, Z. Tang, An adaptive immune response driven by mature, antigen-experienced T and B cells within the microenvironment of oral squamous cell carcinoma, Int. J. Cancer 138 (12) (2016) 2952–2962.
[152] N. Ngamphaiboon, T. Chureemas, T. Siripoon, L. Arsa, N. Trachu, C. Jiarpinitnun, P. Pattaranutaporn, E. Sirachainan, N. Larbcharoensub, Characteristics and impact of programmed death-ligand 1 expression, CD8+ tumor-infiltrating lymphocytes, and p16 status in head and neck squamous cell carcinoma, Med. Oncol. (Northwood, London, England) 36 (2) (2019) 21.
[153] H. Ahn, J.M. Yang, H. Kim, J.H. Chung, S.H. Ahn, W.J. Jeong, J.H. Paik, Clinicopathologic implications of the miR-197/PD-L1 axis in oral squamous cell carcinoma, Oncotarget 8 (39) (2017) 66178.
[154] M. Straub, E. Drecoll, N. Pfarr, W. Weichert, R. Langer, A. Hapfelmeier, C. Götz, K.D. Wolff, A. Kolk, K. Specht, CD274/PD-L1 gene amplification and PD-L1 protein expression are common events in squamous cell carcinoma of the oral cavity, Oncotarget 7 (11) (2016) 12024–12034.
[155] A.K. Mattox, J. Lee, W.H. Westra, R.H. Pierce, R. Ghossein, W.C. Faquin, S.I. Pai, PD-1 expression in head and neck squamous cell carcinomas derives primarily from functionally anergic CD4+ TILs in the presence of PD-L1+ TAMs, Cancer Res. 77 (22) (2017) 6365–6374.
[156] T. Naruse, S. Yanamoto, K. Okuyama, K. Ohmori, H. Tsuchihashi, K. Furukawa, S.I. Yamada, M. Umeda, Immunohistochemical study of PD-1/PD-L1 axis expression in oral tongue squamous cell carcinomas: effect of neoadjuvant chemotherapy on local recurrence, Pathol. Oncol. Res.: POR 26 (2) (2020) 735–742.

[157] Y.M. Lin, W.W. Sung, M.J. Hsieh, S.C. Tsai, H.W. Lai, S.M. Yang, C.J. Chen, High PD-L1 expression correlates with metastasis and poor prognosis in oral squamous cell carcinoma, PLoS One 10 (11) (2015), e0142656.
[158] Y. Maruse, S. Kawano, T. Jinno, R. Matsubara, Y. Goto, N. Kaneko, S. Nakamura, Significant association of increased PD-L1 and PD-1 expression with nodal metastasis and a poor prognosis in oral squamous cell carcinoma, Int. J. Oral Maxillofac. Surg. 47 (7) (2018) 836–845.
[159] T.H. Gasparoto, T.S. de Souza Malaspina, L. Benevides, E.J. de Melo Jr., K.A. Cavassani, J.S. da Silva, A.P. Campanelli, Patients with oral squamous cell carcinoma are characterized by increased frequency of suppressive regulatory T cells in the blood and tumor microenvironment, Cancer Immunology, Immunother.: CII 59 (6) (2010) 819–828.
[160] J.P. Oliveira-Costa, A.F. de Carvalho, D. da Silveira, P. Amaya, Y. Wu, K.J. Park, M.P. Gigliola, M. Lustberg, M.E. Buim, E.N. Ferreira, L.P. Kowalski, J.J. Chalmers, F.A. Soares, D.M. Carraro, A. Ribeiro-Silva, Gene expression patterns through oral squamous cell carcinoma development: PD-L1 expression in primary tumor and circulating tumor cells, Oncotarget 6 (25) (2015) 20902–20920.
[161] A.M. Wirsing, I.K. Ervik, M. Seppola, L. Uhlin-Hansen, S.E. Steigen, E. Hadler-Olsen, Presence of high-endothelial venules correlates with a favorable immune microenvironment in oral squamous cell carcinoma, Mod. Pathol. 31 (6) (2018) 910–922.
[162] G.J. Hanna, S.B. Woo, Y.Y. Li, J.A. Barletta, P.S. Hammerman, J.H. Lorch, Tumor PD-L1 expression is associated with improved survival and lower recurrence risk in young women with oral cavity squamous cell carcinoma, Int. J. Oral Maxillofac. Surg. 47 (5) (2018) 568–577.
[163] H. Takakura, S. Domae, T. Ono, A. Sasaki, The immunological impact of chemotherapy on the tumor microenvironment of oral squamous cell carcinoma, Acta Med. Okayama 71 (3) (2017) 219–226.
[164] O. Kujan, B. van Schaijik, C.S. Farah, Immune checkpoint inhibitors in oral cavity squamous cell carcinoma and oral potentially malignant disorders: a systematic review, Cancer 12 (7) (2020) 1937.
[165] M. Fiedler, F. Weber, M.G. Hautmann, F. Haubner, T.E. Reichert, C. Klingelhöffer, T. Ettl, Biological predictors of radiosensitivity in head and neck squamous cell carcinoma, Clin. Oral Investig. 22 (1) (2018) 189–200.

Index

Note: Page numbers followed by *f* indicate figures and *t* indicate tables.

A

Acetaminophen, 140
Acquired immunodeficiency syndrome (AIDS), 5, 67
Activator protein 1 (AP-1), 89
Adeno-associated viruses (AAVs), 180
Adoptive T-cell chemotherapy (ATC), 179
Age-standardized incidence rates (ASIR), 99
AIDS. *See* Acquired immunodeficiency syndrome (AIDS)
Alphaviruses, 158
American Cancer Society (ACS), 113–114
American Society for Colposcopy and Cervical Pathology (ASCCP), 113–114
Amsel criteria, 87
Anal intraepithelial neoplasia (AIN), 54
Animal models-therapeutic, prophylactic measures, 120*f*, 121–125
  cattle, 122
  dogs, 123
  mus musculus/nude mice, 123–124
  Oryctolagus cuniculus, 121
  Sylvilagus floridanus, 121
  transgenic animals, 124–125
Anisomelic acid, 136
Antigen-presenting cells (APCs), 32
Anti-HIV proteins, 71
Antimycin analogs, 136
Antiproliferative effect, 28
Aptima HPV assay, 112
Arsenic trioxide (As_2O_3), 134
Artemisinin, 140
Artesunate, 140
Aspirin, 140
Ataxia telangiectasia and Rad3-related (ATR) kinase, 26
Ataxia telangiectasia-mutated (ATM) kinase, 26
Atezolizumab (anti-PD-L1), 171–174
ATPase activity, 131
Atypical squamous cells-cannot exclude an HSIL (ASC-H), 105–106
Atypical squamous cells-undetermined significance (ASC-US), 105–106

B

Bortezomib, 135
Bovine papillomavirus (BPV), 122

C

Cadherin-1 (CDH-1), 47
Calixarenes, 137
Cancer, 8
Cancer immunotherapy, 167
Cancer testis antigens (CTAs), 187–188
Canine oral papillomavirus (COPV), 123
Carcinoma in situ (CIS), 108–109
Carrageenan extract (ι-Carrageenan), 137
Catechin, 11
CC. *See* Cervical cancer (CC)
Cellular inhibitor of apoptosis protein 2 (cIAP2), 136
Centers for Disease Control and Prevention (CDC), 113–114
Cervical cancer (CC), 21–22, 65, 99, 119–120, 149, 170–180
  cellular therapies, 179–180
    chimeric antigen receptor (CAR)-T cells, 179–180
    engineered T-cell receptor therapy, 180
  immune checkpoints inhibitors, 170–174, 172–173*t*
    anti-CTLA-4 monoclonal antibodies, 174
    anti-PD1/PD-L1 monoclonal antibodies, 171–174
  therapeutic vaccines, 174–179, 176–177*t*
    genome editing tools, 178–179
    new-generation DNA, 178
    protein, peptide, 175–178
    vector, 175

Cervical intraepithelial neoplasia (CIN), 2, 21–22, 46, 65–66, 150, 178
Cervical scrapings/brushings, 100
Cervicovaginal microbiota (CVM), 82*f*, 83–86
  classification, 84
  preventive role, CC, 85–86
  reproductive health, 84–85
  restoration, CC prevention, 90
  sexually transmitted infections (STIs)and CC, 86–90
    bacterial vaginosis (BV), 87
    Chlamydia trachomatis, 88–89
    HPV infection, 86–87
    human immunodeficiency virus (HIV), 89–90
    Trichomonas vaginalis, 87–88
Cervista HPV HR assay, 111–112
Chlamydia trachomatis (CT), 84–85
Cidofovir, 138
CIN. *See* Cervical intraepithelial neoplasia (CIN)
Clinical manifestation, of HPV infections, 7–10, 7–8*f*
  anal cancer, 9–10
  cervical cancer (CC), 7–8
  mouth, throat cancer, 10
  penile cancer, 9
  vaginal cancer, 8–9
  vulvar cancer, 9
COBAS HPV test, 110–111
Colony-forming units (CFUs), 175
Colposcopy, 105–106
Community state types (CST), 84
Co-testing, 110
Cottontail rabbit papillomavirus (CRPV), 121
CTAs. *See* Cancer testis antigens (CTAs)
Curcumin, 135–136
Cyclic GMP-AMP synthase stimulator of interferon genes (cGAS-STING), 184–185
Cyclic guanosine monophosphate (GMP)-adenosine monophosphate (AMP) synthase (cGAS), 30–31
Cyclin-dependent kinase (CDK) inhibitors, 109
Cyclooxygenase 2 (COX-2) enzyme, 140
Cytomegalovirus (CMV), 138
Cytotoxic T-lymphocyte-associated protein 4 (CTLA-4), 170–171
Cytotoxic T lymphocytes (CTLs), 32, 74–75

D

Defensins (antimicrobial peptide), 71
Dendritic cells (DCs), 68, 156
Diagnostic techniques, 100–112
  cytology/Papanicolaou test/Pap smear, 104–108, 106–107*f*
  histology, 108–110, 109*t*
  nucleic acid-based detection, 110–112
  sampling strategies, 100–104, 102–103*t*
Diallyl disulfide (DADS), 136
Diallyl sulfide (DAS), 136
Differentiated vulvar intraepithelial neoplasia (dVIN), 50
Digene hybrid capture 2 HPV DNA test (HC 2 Qiagen), 111
Dihydrotanshinone I (DI), 132–133
Doggybone (db) DNA, 185–186
Double-stranded DNA (dsDNA), 129
Double-stranded RNA (dsRNA), 135

E

E6-associated binding protein (E6AP), 24–26, 47
E2F transcription factor 1 (E2F1), 24–26
E2F transcription factor 6 (E2F6), 24–26
Electroporation (EP), 184–185
Endoplasmic reticulum and Golgi (ER-GC) complex, 31
Epidermal growth factor (EGF), 23
Epidermal growth factor receptors (EGFRs), 23
Epigallocatechin-3-gallate (EGCG), 136
Esophageal adenocarcinoma (EAC), 53
Estrogen receptor α (ERα), 136
Eukaryotic translation initiation factor 4E (eIF4E), 134
European Society of Medical Oncology (ESMO), 182–183
Existing drugs, repurposing, 137–140, 139*t*
  antibiotic, 138–140
  antiinflammatory, 140

antiparasitic, 140
antiviral, 138
Extracellular signal-regulated kinase (ERK), 24–26

F

Fluorescence resonance energy transfer (FRET), 111–112
Foreign agents (antigens), 150–151

G

Gamma delta 1 (GD1) cells, 89
Gamma-retrovirus, 180
Gastroesophageal reflux disease (GERD), 53
Granulocyte-macrophage colony-stimulating factor (GM-CSF), 29, 159
Gynecologic oncology group (GOG), 170

H

HAART. *See* Highly active antiretroviral therapy (HAART)
Head and neck squamous cell carcinoma(s) (HNSCC(s)), 48–49, 181–189
immune checkpoint inhibitors, 181–184
nivolumab, 183–184
pembrolizumab, 182–183
novel immunotherapeutic strategies, 188–189
systemic cell-mediated immunotherapy, 188–189
targeted immunotherapy, 189
therapeutic vaccines, 184–188
DNA-based, 184–186
mRNA, 186
peptide-based, 187–188
Heparin sulfate proteoglycans (HSPGs), 23
High-grade squamous intraepithelial lesion(s) (HSIL(s)), 48, 68, 82–83, 105–106
Highly active antiretroviral therapy (HAART), 72
High-risk-human papillomavirus (HR-HPV), 1, 21–22, 45, 65, 99–100, 119–120
Histocompatibility leukocyte antigen (HLA), 65–66
HIV. *See* Human immunodeficiency virus (HIV)
HIV-inducing factor (HIF), 89
Host immune evasion, 30–32
Host immune system, 150–151
HPV. *See* Human papillomavirus (HPV)
HPV-HIV co-infections, gynecological disorders, 72–75, 73*t*
anogenital warts, vulvar intraepithelial neoplasia, 75
CC pathogenesis, 73–75, 74*f*
HPV-specific T lymphocytes (HPVST), 179
HPV structure, replication, pathogenesis, 22–27, 22*f*, 24*f*
Human immunodeficiency virus (HIV), 12–13, 67
Human leukocyte antigen (HLA), 156–157
Human papillomavirus (HPV), 1, 21–22, 43, 65, 82–83, 119, 129, 149, 150*f*
cancers associated
immunotherapies, 169–170
therapeutic options, 168–169
cervical cancer (CC), 83
related malignancies, 166*f*, 167, 168*f*
types, 82–83
Hybridization protection assay (HPA), 112

I

IFN. *See* Interferon (IFN)
Immune response, 27–30
adaptive immunity, 29–30
innate immunity, 27–29
Immunological aspects
HPV infection, 68–70
HPV on HIV infection, 70–72
Immunological escape strategy, 43–45
Inflammation, 140
Interferon (IFN), 2, 69
Interferon regulatory factor1 (IRF1), 24–26
Interferon regulatory transcription factors, 31
Interleukin 7 (IL-7), 178
Invasive cervical carcinoma (ICC), 87
Irradiation (IR), 132–133

K

Keratinocytes (KCs), 68
Koilocytosis, 104–105

L

Lactobacillus, 85–86
Langerhans cells (LC), 68
Lentivirus, 180
Liquid-based cytology (LBC), 101
Listeria monocytogenes (LM), 175
Liver kinase B1 (LKB1), 47
Long control region (LCR), 22
Loop electrosurgical excision procedure/process (LEEP), 55, 114–115
Lower anogenital squamous terminology (LAST), 109
Low-grade squamous intraepithelial lesion (LSIL), 46, 105–106
Low-risk human papillomavirus (LR-HPV), 1, 21–22, 45, 65, 82–83, 99–100, 123

M

Major histocompatibility complex (MHC), 156
Major histocompatibility complex I (MHC I), 21–22
Malignancies, epidemiology, pathology, 7, 44–45*f*, 46–55
- anal cancer, 54
- cervical cancer (CC), 46–48
- esophageal cancer/esophageal squamous cell carcinoma (ESCC), 53
- genital warts/anogenital warts/condyloma acuminatum, 54–55
- oropharyngeal cancer (OPC), 48–50
- penile cancer, 50–53, 52*f*
- squamous cell, 44–45*f*, 46–55
- treatment, 55–56
- vaginal cancer, 53–54
- vulvar cancer/vulvar squamous cell carcinoma (VSCC), 50

Map kinase (MAPK), 24–26
Matrix metalloproteinases-2 (MMP2), 85–86
Microbiome, 83
Modified Reid colposcopic index (RCI), 101
Modified vaccinia Ankara (MVA) vector-based vaccine, 156
Modulated retinoblastoma protein (pRB), 134
Mouse papillomavirus (MmuPV), 119–120
Mus musculus, 123–124

N

Natural killer cells (NKCs), 68
Nonsteroidal antiinflammatory drugs (NSAIDs), 140
Nuclear factor kappa B (NF-κB), 27–28, 89
Nucleic acid amplification technology (NAAT), 110
Nucleic acid sequence-based amplification (NASBA), 110

O

Onclarity HPV test, 111
Open reading frames (ORFs), 3, 23–24
Oral cancer, 190–191
- immune checkpoint inhibitors, 190–191

Oral lichen planus (OLP), 190–191
Oral potentially malignant disorders (OPMDs), 190–191

P

Papanicolaou test/Pap smear test, 65–66
Papillomavirus (PV), 119–120, 129
Pathogenesis, immune response, 22–30
Pathogen recognition receptors (PRRs), 30
Penile intraepithelial neoplasia (PIN), 46
Peripheral blood mononuclear cells (PBMCs), 179–180
Pillararenes, 137
Podophyllin, 11
Polymerase chain reaction (PCR), 6, 110
Potential drug targets, 130–137
- E1, E2 proteins, 130–131, 133*f*
- E6 oncoprotein, 132–135
- E7 oncoprotein, 135–137
- E5 protein, 131–132
- L1 capsid protein, 137

Primary HPV screening tests, 110
Programmed cell death 4 (PDCD4), 47
Programmed death ligand 1 (PD-L1), 169
Programmed death receptor 1 (PD-1), 170–171
Promyelocytic leukemia (PML), 23
Prophylactic vaccines, 151

Prostate-specific membrane antigen (PSMA), 179–180
Protein tyrosine phosphatase nonreceptor type 13 (PTPN13), 135–136

Q
Quasi-enveloped double-stranded DNA (dsDNA) virus, 21–22
Quercetin, 133–134

R
Radiation therapy (RT), 48
Rapid amplification of cDNA ends (RACE), 180
Rapid capture system, 111
Recombinant/chimeric DNA (rDNA) technology, 160
Recombinant human granulocyte-macrophage colony-stimulating factor (rhGM-CSF), 171–174
Recurrent respiratory papillomatosis (RRP), 119–120, 149–150
Regulated-on-activation normal T-expressed and secreted factor (RANTES), 70–71
Regulatory T cells (Tregs), 32
Replication cycle, 3–4, 23

S
Screening strategies, 112–115
Self-sampling kits, 104
Semliki Forest virus, 158
Sexually transmitted infection(s) (STI(s)), 1–2, 149–150
Signal transducer and activator of transcription 1 (STAT1), 24–26
Sindbis virus, 158
Single chain of variable fragment (scFv), 179–180
Single-nucleotide polymorphisms (SNPs), 174
Small interfering RNA (siRNA), 135, 159
Squamous cell carcinoma (SCC), 43, 105–106
Squamous intraepithelial lesions (SIL), 29–30
Stable disease (SD), 182–183
Suicidal DNA, 158
Surface layer proteins (SLPs), 175–178
Symptoms, 5
  common/flat/plantar warts, 5
  genital warts, 5
Synthetic long peptides (SLPs), 184
Synthetic macrocycles, 137

T
Tanshinone IIA (Tan IIA), 134
TATA-binding protein (TBP), 24–26
T-cell receptors (TCRs), 180
T helper cells (Th)-1, 70–71
Therapeutic vaccines, 152
Thrombospondins, 71
TLR. *See* Toll-like receptor (TLR)
T lymphocytes, 68
TNF-receptor-associated factor (TRAF), 31
Toll-like receptor 9 (TLR9), 21–22, 30
Toll-like receptor (TLR), 65–66, 175–178
Transcription activator-like effector nuclease (TALENs), 178–179
Transcription-mediated amplification (TMA), 110
Transforming growth factor-β (TGF-β), 2, 27, 69
Trichomoniasis, 87–88
Tumor-associated antigens (TAAs), 179–180
Tumor-infiltrating lymphocytes (TILs), 179
Tumor necrosis factor (TNF), 2, 27, 69
Tumor-specific mutated proteins, 189
Tyrosine kinase 2 (TYK2), 31

U
Ubiquitin C-terminal hydrolase L1 (UCHL1), 31
Upstream regulatory region (URR), 3
US Food and Drug Administration (FDA), 169
U1 small nuclear ribonucleoprotein (U1snRNP), 26–27
US preventive services task force (USPSTF), 5–6
Usual-type vulvar intraepithelial neoplasia (uVIN), 50

V

Vaccine, development, 11–12, 153–160
  candidate, 152–153
    E6/E7 oncoproteins, 153
    VLPs prophylactic, 152
  chimeric VLP, 155–156
  epitope-based, 159
  L2-based prophylactic, 155
  L1 capsomere-based, 153–155
  licensed, 153, 154*t*
  nucleic acid-based, 157–158
    DNA, 157
    RNA replicon, 158
  protein/peptide-based, 156–157
  type, requirement, 151–152
    prophylactic, 151
    therapeutic, 152
  vector-based, 156
  whole cell-based, 158–159
    DC, 158–159
    tumor cell, 159
Vaccinia virus, 156
Vaginal dysbiosis, 84–85
Vaginal eubiosis, 85–86
Vaginal intraepithelial neoplasia (VAIN), 46
Vascular endothelial growth factor (VEGF), 69–70, 135
Venezuelan equine encephalitis virus (VEE), 158
VIN. *See* Vulval/vulvar intraepithelial neoplasia (VIN)
Virus-like particles (VLPs), 22–23
Visual examination under Lugol's iodine (VILI), 114–115
Visual inspection with acetic acid (VIA), 114–115
Vulval/vulvar intraepithelial neoplasia (VIN), 46, 75, 150
Vulvar non-neoplastic epithelial disorders (VNEDs), 50
Vulvovaginal candidiasis (VVC), 84–85

W

Wart treatment, 11
  chemical, 11
  physical, 11
Withaferin A (WA), 133
World Health Organization (WHO), 1, 45, 100

Z

Zinc finger nucleases (ZFNs), 178–179

Kolibri 360 GmbH
Ferdinand-Jühlke-Straße 7
99095 Erfurt, Deutschland
produktsicherheit@kolibri360.de